MIND
BLOWN

MIND BLOWN

DAN MARSHALL

LOST
THE
PLOT

LOST THE PLOT

A Lost the Plot book, first published in 2019 by Pantera Press Pty Limited
www.PanteraPress.com

Please send all permission queries to:
Pantera Press, P.O. Box 1989, Neutral Bay, NSW, Australia 2089 or info@PanteraPress.com

A Cataloguing-in-Publication entry for this book is available from the National Library of Australia.
ISBN 978-1-925700-86-2 (Hardback)

Cover, internal design and layout: Dan Marshall
Publisher: Martin Green
Editors: Anne Reilly and Anna Blackie
Proofreader: Cristina Briones
Printed and bound in China by Shenzhen Jinhao Colour Printing

Pantera Press policy is to use papers that are natural, renewable and recyclable products made from wood grown in sustainable forests. The logging and manufacturing processes are expected to conform to the environmental regulations of the country of origin.

To Holly, Milly and Winnie.
All three of you blow my mind.
xx

Introduction.

As you start to read this book, have a good look around you. Chances are, things seem relatively sedate, your surroundings calm. There'll be no sense at all that you're actually on the surface of a giant spherical rock that's rotating at 1600 km/hour as it hurtles at 110,000 km/hour through near infinite space, circling a giant sphere of burning plasma. Or that our giant spherical rock — along with others of its ilk and several humongous balls of gas — is part of a solar system that's moving at an average speed of 828,000 km/hour as it travels around the centre of our Milky Way galaxy.

It isn't only these speeds that are astonishing, it's the volumes too. Our solar system is just one of possibly 400 billion such systems in our galaxy. What's more, our galaxy is just one of possibly 500 billion galaxies in the observable universe.

As your brain attempts to process these colossal cosmic numbers, it will use its 100 billion neurons to do so. Give or take a few billion.

I remember my first moment of cosmic realisation – and the accompanying awareness that our lives, and the universe we live in, are not quite as they initially seemed. It was electrifying. I could feel my mind being well and truly blown. And since then, I've had that experience many times.

This book is a carefully curated collection of facts with accompanying illustrations that aims to demonstrate just how spectacularly strange the universe around us is. From quantum mechanics and general relativity to immortal jellyfish and mathematical insects, this book will take you from the origins of the universe and life here on Earth, to everywhere in between.

Whilst every effort has been made to ensure all the mind-blowing facts in the book are as up to date as possible, such is the wonderfully changeable nature of our universe, that some things may have altered slightly since it was first published.

As the eminent British-Indian scientist J.B.S. Haldane famously declared in his essay collection *Possible Worlds* (Harper & Brothers, New York, 1928):

'My own suspicion is that the universe is not only queerer than we suppose, but queerer than we can suppose.'

He was dead right too.

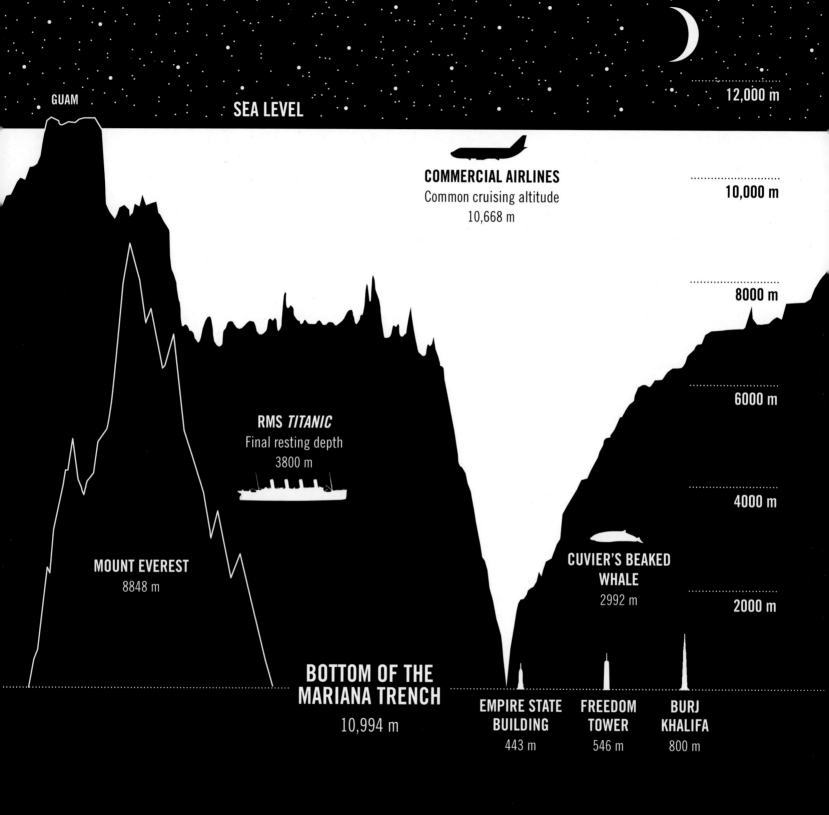

GUAM

SEA LEVEL

12,000 m

COMMERCIAL AIRLINES
Common cruising altitude
10,668 m

10,000 m

8000 m

RMS TITANIC
Final resting depth
3800 m

6000 m

4000 m

CUVIER'S BEAKED WHALE
2992 m

MOUNT EVEREST
8848 m

2000 m

BOTTOM OF THE MARIANA TRENCH
10,994 m

EMPIRE STATE BUILDING
443 m

FREEDOM TOWER
546 m

BURJ KHALIFA
800 m

While 12 people have walked on the Moon, only four have been to the deepest natural point on Earth.

It's funny how we know more about an orbiting celestial body 384,400 kilometres away than we do about the deepest parts of our oceans here on Earth.

From our limited knowledge, we've managed to create some ocean floor maps, but even then we've only managed to look at less than 10% of the ocean floor. Considering Earth's surface is 70% ocean, that leaves an awful lot of our planet unexplored. The least understood biological habitat on Earth happens to be its largest. Go figure!

By contrast the National Aeronautics and Space Administration (NASA) in the United States recently released the highest resolution near-global topographic map of the Moon ever created. Famously, it included its lesser known dark side, which is not in our line of sight from Earth. We are familiar with all the Moon's major geological terrains, the impact craters and its volcanic features.

Of the 12 people to set foot on the surface of the Moon to date, all were astronauts from the Apollo program, which NASA ran from 1961 to 1972. Neil Armstrong and Edwin 'Buzz' Aldrin of Apollo 11, were the first to land. Over the next three years, NASA managed another five successful excursions to the lunar surface.

Back here on Earth, the deepest natural point on its surface is at the bottom of the Mariana Trench in the western Pacific Ocean. The maximum known depth is 10,994 metres. To give that some context, if you were to drop Mount Everest into the trench, it would comfortably fit with two kilometres to spare above its peak.

So far, only four people have been to the bottom of the Mariana Trench. In 1960, oceanographers Don Walsh and Jacques Piccard descended in their submersible. In 2012, Canadian film director James Cameron travelled to the bottom and in 2019 businesssman Victor Vescovo undertook the same incredible journey.

When you shuffle a deck of cards, that exact order has never been seen before in the history of the universe.

There are 52 playing cards in a deck. That's not really a number to get too excited about. But, if you were to give that deck a shuffle and lay those 52 cards down in a line, things would quickly move into the mind-blowing zone. The order of cards in front of you has never been seen before. Ever.

If you look at the maths behind it, a deck of 52 cards can be ordered like this: 52 x 51 x 50 x 49 x ... x 3 x 2 x 1. To express this in words, there are 52 ways to choose the first card, 51 ways to choose the second, 50 ways to choose the third, and so on.

The number of order possibilities that result from that maths is: 80,658,175,170,943,878,571,660,636, 856,403,766,975,289,505,440,883, 277,824,000,000,000,000.

Or, to use the concise form — if ridiculously large numbers make your head fall off — 8×10^{67}.

This number is big. Bigger than astronomical. So big that it even goes beyond cosmic. This is a number bigger than the total number of all the stars in the universe!

This number of possibilities is so massive that if someone had been shuffling a deck of cards once per second since the very beginning of the universe — over 14 *billion* years ago — they wouldn't have even shuffled the deck 10^{18} times.

With this beyond cosmic number of possibilities in mind, it's a safe bet that any order of cards drawn through random shuffling is likely to never have appeared before — and to never appear again in your lifetime!

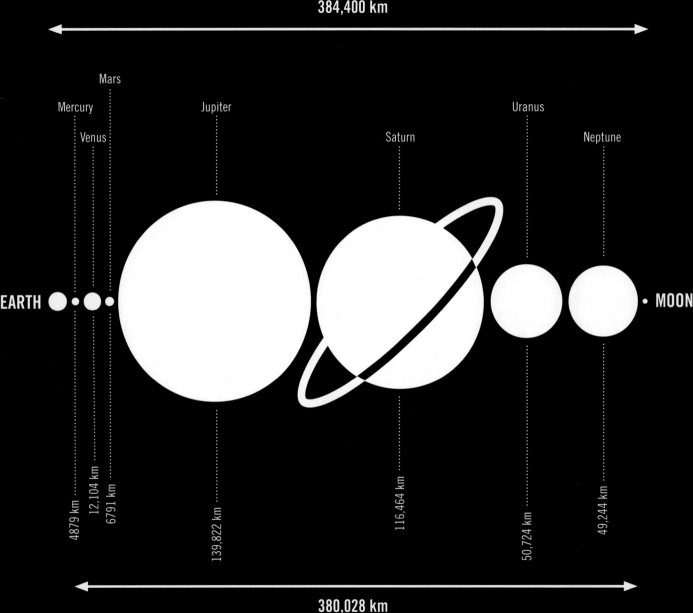

384,400 km

Mercury

Venus

Mars

Jupiter

Saturn

Uranus

Neptune

EARTH

MOON

4879 km

12,104 km

6791 km

139,822 km

116,464 km

50,724 km

49,244 km

380,028 km

All the planets in our solar system can fit in the space between the Earth and Moon.

Nobody knows who first used the word 'space' to refer to the depths of the cosmos … What we do know is that whoever it was absolutely nailed it — our universe is mind-blowingly massive.

The Moon is the Earth's nearest celestial neighbour, and our only permanent natural satellite. We can look up to the sky every night (and some days) and there it is, locked in orbit with us.

Don't be fooled into thinking that, just because we can see it, the Moon is literally nearby. This neighbour lives an enormously long way from us. The total distance from Earth to the Moon is a jaw-dropping 384,400 kilometres. It's so far away that when Apollo 11 first journeyed there, it took almost three days to arrive at the Moon. Travelling the same distance, you could do almost ten full circles of the Earth!

Understanding just how huge our universe is, or the sizes of all of the weird and wonderful things that fill it, is astronomically hard to do.

If you added together the diameters of all the planets in our solar system — as per NASA's measurements — they would equal 380,028 kilometres. That massive number is smaller than the distance from Earth to the Moon! This means that all the planets could fit in the space between Earth and the Moon, with room to spare — 4372 kilometres, to be precise.

Of course, Pluto is once again counted as a planet — albeit with dwarf status — and those figures don't include it. But it too could squeeze in, with its teeny 2300 kilometres of width, and the planets still wouldn't need to bump up against each other!

The largest animal to ever exist can't swallow anything larger than a grapefruit.

Blue whales are the undisputed kings of the oceans. Three times the size of the largest dinosaurs of the land and sea, and bigger than any other creature alive today.

You'd think that the creature holding the title of 'Largest Animal to Ever Swim the Oceans' would be feasting on food that's correspondingly enormous. You would, however, be wrong. Very wrong. The blue whale's throat is teeny, which stops it from devouring anything larger than a grapefruit. But when a blue whale eats, it definitely eats in quantity!

This whale's food of choice is krill. These minute pink shrimp-like creatures slip down the whale's slim gullet with remarkable ease. Plus, these whale snacks swarm in belly-filling numbers, with some mobs weighing in at a hefty 90 million kilograms — roughly equivalent to 643 blue whales!

Sometimes, there are extended gaps between meals. When travelling and breeding, blue whales may endure six to eight months of near starvation. As you'd expect, when back in waters where krill are plentiful, peckish blue whales don't hold back. During their feeding season, they can gorge on more than 40 million krill in a single day. That's around 4000 kilograms of food! Staggering, considering blue whales weigh 140,000 kilograms on average.

To chow down on such monstrous hordes of krill, blue whales swim with their massive mouths wide open, swallowing their gargantuan shrimp cocktail whole. In the process, they also take in huge amounts of water, which they filter out with their ginormous 2700-kilogram Jabba the Hut-like tongues.

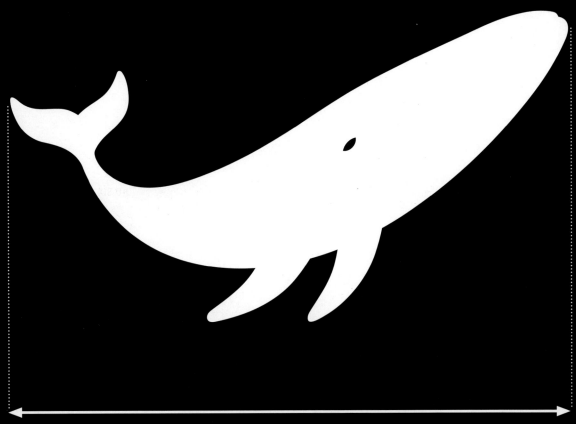

2500 cm

15 cm

The world's nine richest men have more wealth than half of all people on Earth.

It's more than slightly disturbing to realise that nine blokes are collectively wealthier than the 3.85 billion people who count as the poorest half of humanity. The nine billionaires who make up this list are:

Jeff Bezos, the American founder, chairman and chief executive of Amazon, who as of April 2019 had an estimated net worth of US$115 billion.

Bill Gates, the founder of Microsoft, with a fat stack of US$102 billion.

Warren Buffett, the American CEO and largest shareholder in Berkshire Hathaway, the third richest man, with US$82.9 billion.

Bernard Arnault, the French CEO of LVMH — the world's leading luxury products group — who has an enviable net worth of US$80.5 billion.

Amancio Ortega, Spanish founder of Inditex, which owns the Zara fashion chain, is worth approximately US$65.7 billion.

Larry Ellison, the American co-founder and CEO of Oracle Corporation, whose wallet is bulging with his US$63.9 billion.

Carlos Slim Helú, the Mexican owner of global conglomerate company Grupo Carso, who just missed out on the title of 'World's Richest Man Number 8' with his US$62.7 billion.

Mark Zuckerberg, the American chairman, CEO and co-founder of Facebook. Mark's endeavours have netted him a cool US$60.3 billion.

And, finally, **Larry Page**: the American co-founder and CEO of Google who has a comparatively humble fortune, with an estimated net worth of US$52 billion.

The net worth of these nine men currently exceeds US$700 billion, though this number is subject to sudden change. If these billionaires continue to see high returns on their incomprehensible wealth, we could see the world's first trillionaire in as little as 25 years.

The atomic bomb was built and detonated 20 years before we had colour television.

Atoms are minuscule things, impossible to see with the naked eye. But if you force one into parts, you set off an explosive chain reaction that generates a blast so deadly and immense that it would eviscerate you and everything around you in less than a second. Boom!

Scientist Ernest Rutherford made history in 1917 by splitting the atom split the atom, in a nuclear reaction now known as fission. By 1939, the Allied forces of the Second World War (Great Britain, France, the Soviet Union, the United States and China) were discussing how nuclear fission might be used for military purposes, and plans for the creation of a uranium bomb were made.

That same year, Albert Einstein wrote to US President Franklin Roosevelt to give scientific support to the theory behind the atomic bomb. He also expressed concern that the Nazis were working on a powerful new weapon of their own.

Consequently, the Manhattan Project was set up. In 1943, theory and practice came together when construction of a deliverable atomic bomb was completed.

Then came tests. In 1945 it was detonated in the darkness of the New Mexico desert. Onlookers watched as an enormous mushroom cloud of searing light exploded 12 kilometres into the air, powered by 18 million kilograms of TNT. The explosion illuminated the surrounding mountains brighter than daytime for one to two seconds. Subsequently, the bombs were used on Japan to catastrophic effect.

While these incredible leaps in military science were being made, television lagged behind. Amazingly, it wasn't until 20 years later that colour television sets started selling widely. The first all-colour prime-time season came in 1966, when viewers tuned in to the short-lived American sitcom *The Marriage*. The era of black and white TV was over.

Fleas can jump more than 200 times their body length and accelerate faster than a rocket.

Don't you wish fleas were more discriminating! These parasites are happy to feed off the blood of our beloved pets and just as happy to feed off us if the need arises.

At best you'd call them a nuisance, but it's probably time to cut them some slack. There's way more to fleas than you might realise. Relative to their body size, they're some of the most extraordinary high jumpers in the world!

Because they're so small, in order to find an animal they can feed off, they have to be able to leap, and they sure can leap … These tiny creatures can jump over 200 times their own body length — in human terms, that would be like jumping over the Eiffel Tower!

It's not only their jumping ability that's incredible, but their powerful acceleration. Relative to their size, they accelerate faster than a rocket lifting off for space — 50 times faster! Taking into account their size and the force used to jump those great distances, a flea's acceleration is approximately 100 times the force of gravity. Compare that with fighter pilots, who can generally only handle acceleration up to nine times the force of gravity before passing out.

As well as being amazing at jumping, accelerating and dealing with g-force, fleas are strong too. They can lift objects 150 times heavier than their own body weight! To give this some context, the average human can just about lift an object identical to their own body weight.

The International Space Station is the most expensive object ever built.

Two things humans are extremely good at are spending money and building things. When we spend massive amounts of cash to build incredible things, the results can be literally out of this world. In the case of the International Space Station – or ISS – we're talking 408 kilometres out of this world, to be exact. The ISS travels at a mind-blowing 27,600 km/hour in lower Earth orbit.

Sixteen nations were involved in the station's construction: the United States, Russia, Canada, Japan, Belgium, Brazil, Denmark, France, Germany, Italy, the Netherlands, Norway, Spain, Sweden, Switzerland and the United Kingdom. But it was America that generously picked up the bulk of the station's US$150 billion (and counting) construction cost, paying an astronomical US$100 billion to date.

The ISS is so massive that it would have been impossible to build on the ground and then launch into space in one go. There isn't a rocket big or powerful enough to handle its 420,000-kilogram load. It's a real beast! The solution to this problem was to take the station up into space piece by piece and gradually build it in orbit. The first piece went up in 1998, and construction continues to this day.

As the most expensive and ambitious structure humans have ever assembled, the ISS has been a complete success. It should give us a warm and fuzzy feeling that when we all work together, we can achieve wonders!

The ISS is scheduled for retirement in 2024, but for the time being it remains clearly visible in the night sky.

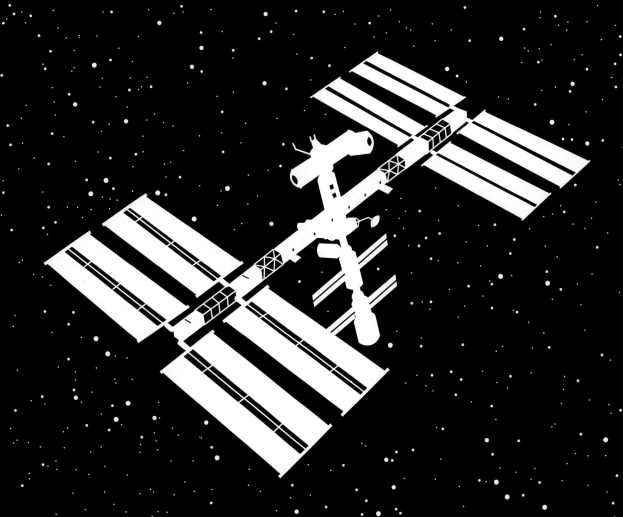

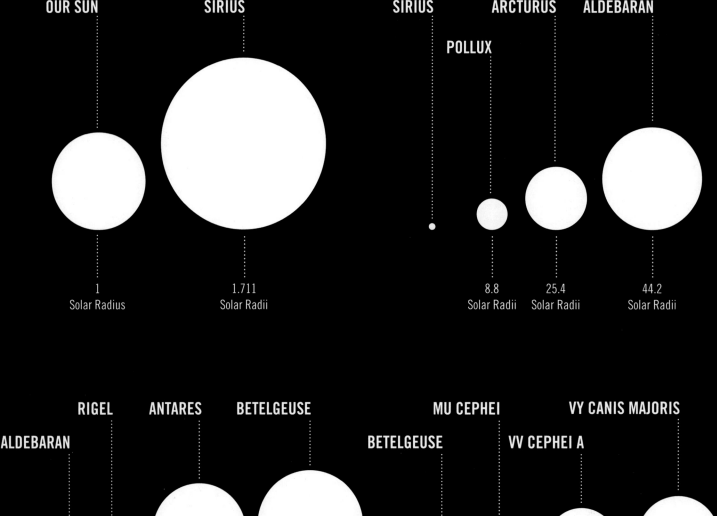

OUR SUN
1
Solar Radius

SIRIUS
1.711
Solar Radii

SIRIUS

POLLUX
8.8
Solar Radii

ARCTURUS
25.4
Solar Radii

ALDEBARAN
44.2
Solar Radii

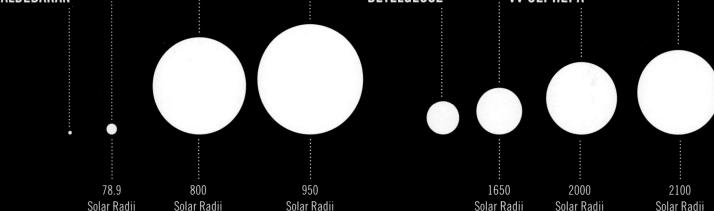

ALDEBARAN

RIGEL
78.9
Solar Radii

ANTARES
800
Solar Radii

BETELGEUSE
950
Solar Radii

BETELGEUSE

MU CEPHEI
1650
Solar Radii

VV CEPHEI A

VY CANIS MAJORIS
2000
Solar Radii

2100
Solar Radii

Some stars in our galaxy are thousands of times bigger than our Sun.

Our Sun may appear to be the biggest thing in the sky, but that's only because of how close it is to us. This is why it looks so much larger and brighter than all the other stars we can see. But that yellow dwarf is pulling a bit of a fast one on us …

Don't be fooled by how tiny those other dots of light may look. They're actually enormous balls of burning gas, all of which are many times greater and brighter than our own Sun. Our Sun could be classified as an average-sized star. Although some stars are smaller, some are most definitely bigger. Oodles bigger.

The largest we've detected so far is VY Canis Majoris, which is located 3900 light years from Earth. Its radius is roughly 2000 times that of the Sun. It would take a passenger jet 1100 years to fly round it once!

VY Canis Majoris is what's known as a red hypergiant. These are classified by their enormous mass and luminosity. They burn fuel very quickly, so they only last a few million years, which in cosmic terms isn't long at all. By way of comparison, the Sun will generate nuclear fusion for eight billion years!

We expect VY Canis Majoris to die and go supernova within the next 100,000 years. When it does, it will explode and seed the universe with heavy elements essential for life on Earth, at the same time forming a black hole.

Despite its mind-blowing size, from Earth VY Canis Majoris looks like nothing but the merest speck buried among the billions of other stars in the Milky Way. Beyond that lie hundreds of billions more galaxies, and some may have stars that are even bigger still.

99.9% of all life that has ever existed on Earth is now extinct.

Life loves Earth. Not only do we see an abundance of it across the globe today, but palaeontologists are kept busy by a steady stream of new discoveries of species from long ago. Unearthing dinosaurs would be hard to top, but palaeontologists keep trying. And they'll never run out of work because the number of species currently alive is a fraction of the number that have previously evolved, lived and then become extinct.

Palaeontologists can tell that most species tend to hang around for anything from one million to 11 million years. We modern humans have been around for 200,000 years.

But old age isn't the only reason species die out. Disconcertingly, there have been five mass extinctions in the history of Earth. The first was 444 million years ago, when 86% of all species died as the planet froze over. Then 375 million years ago, 75% of all life suffocated as oxygen was sucked from the water, possibly by algal blooms. The biggest extinction of all was the third, around 250 million years ago.

A mind-blowing 96% of all life on Earth was extinguished as global temperatures surged and the oceans acidified and stagnated. Definitely not a fun time to claim Earth as your home! The enigmatic fourth extinction was 200 million years ago: no clear cause has yet been discovered.

The mass extinction that devastated the dinosaurs was the most recent, only 65 million years ago. Scientists are pretty certain that a gobsmackingly large meteorite hit the Earth and killed off 88% of life.

Could we be in the midst of the sixth? Scientists believe that 150–200 species of plants, insects, birds and mammals are dying out every single day. This extinction rate is themost rapid and dramatic the world has experienced since the dinosaurs were wiped out.

It's pretty crazy to realise that human activity is now directly influencing extinction. We've witnessed the disappearance of 60% of animal life since 1970. The world's environmental experts are warning that the annihilation of wildlife is an emergency that threatens our entire civilisation and the future of our planet.

5614 km

5567 km

There is a fence in Australia that is the equivalent distance of London to New York.

Australia is big. Very big. In fact, it's enormous. So enormous that in South-East Australia there's a fence that stretches the equivalent distance of New York to London. That's an almost 6000 km stretch of fencing, and it's all to keep dingoes at bay.

The imaginatively titled 'Dingo Fence' is made of wire-mesh and timber posts – stretching for 5614 km, separating the dingoes from the fertile land where sheep and cattle graze.

Be careful leaning on it, though. This fence is partially electrified, and costs a million dollars a year to maintain – that's enough to fly from London to New York nearly 2000 times! At night, the fence is illuminated by red and white fluorescent lamps and patrolled by 23 full-time employees.

The Dingo Fence extends all the way from Brisbane down to the Nullarbor Plain in South Australia. If you wanted to walk it, you would need to clear a full six months in your calendar and invest in a good pair of walking shoes.

Mind-blowing distances are pretty common to the Australian landscape. After all, Australia is the planet's sixth largest country after Russia, Canada, China, the United States and Brazil.

Anna Creek in South Australia oversees 6,000,000 acres, a land-size that is larger than Israel! More than 17,000 heads of cattle are reared there, which makes it the biggest beef cattle producer in the world.

If there was a hole right through the Earth, it would take 42 minutes to go from one side to the other.

If you've ever wondered just how far we humans have drilled down into our planet's surface, then try imagining the globe as an apple. It might surprise you that we haven't even sunk our teeth through the skin yet. Bizarrely, some people from the Soviet Union once started digging to see how deep a hole they could make. After almost two decades of constant work, they'd only managed to go down 12 kilometres – that's around 0.1% of the way through the planet.

Creating a hole all the way through Earth would be impossible. The tunnel itself would be 12,756 kilometres long, and all of the material in its path would need to be displaced. That would make a very impressive pile. Not only would the tunnel be long but inside it would be hot as all hell. The temperature at Earth's core is a blazing 6000°C!

Imagine for one minute that the Soviets had succeeded in digging their crazy hole. Somehow that hole would go all the way through Earth, ignoring the enormous distance and the heat, not to mention other barriers, like the friction of the drilling process and the rotation of Earth itself.

What would happen if someone was to fall through it? Theoretically, you'd enter the tunnel at one point, fall straight through the centre and then resurface on the opposite side of the planet. Easy peasy. You'd start falling with zero speed, which would rapidly build to a maximum of 28,440 km/hour as you hit the Earth's core. You'd then slow back to zero as you reached the other side. Much like the way a pendulum swings down and up again.

The total travel time for this theoretical journey from one side of the globe to the other would be a tidy 42 minutes. That's less time than the average television drama episode.

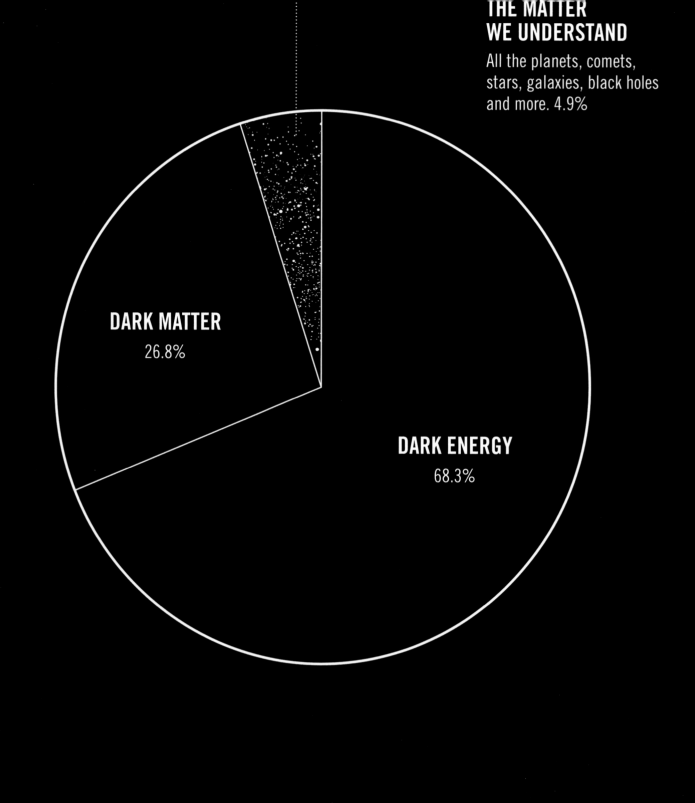

THE MATTER WE UNDERSTAND

All the planets, comets, stars, galaxies, black holes and more. 4.9%

DARK MATTER
26.8%

DARK ENERGY
68.3%

95.1% of the universe is made of stuff we can't see, detect or even comprehend.

It's bonkers, but for all of the amazing discoveries made and the incredible depth of knowledge mankind has accumulated, we still know very little of the universe in which we live. For instance, astronomers have calculated that the stars, planets and galaxies we can see today – which we call the observable universe – make up just 4.9% of the entire cosmos. The other 95.1% is a mystery to us, although a lot of effort goes into studying it.

From what we can tell, it appears to be made of an invisible substance we've named dark matter (26.8%) and a force that repels gravity called dark energy (68.3%).

We've never been able to observe dark matter. Its inability to interact with matter makes it completely invisible and impossible to detect with current instruments. We know dark matter exerts a gravitational effect on galaxies and galaxy clusters.

This causes stars to orbit at more or less the same speed, regardless of where they are in a galaxy, rather than slowing down at the edges and speeding up near the centre as you would expect.

It's also known that the universe is still expanding, even though the Big Bang – from which the universe was born – happened billions of years ago. You'd assume that, with gravity, that expansion would have slowed. But, mind-blowingly, instead it's speeding up. That's because dark energy is increasing in strength as the universe expands.

It's astonishing that so much of our universe consists of forces we can barely comprehend, let alone see.

Most people who stutter are fluent when singing, reading aloud to themselves or talking to their pets.

Stuttering is a disorder that affects the fluency of speech. Stutterers know exactly what it is they want to say; they just have trouble saying it. The speech of a stutterer is generally disrupted by repeating sounds, words or phrases; by prolonging sounds; or having moments where no sound comes out at all. In some cases, people who stutter also experience physical disruptions like head movements, irregular blinking and facial twitches and tremors.

We think that stuttering is related to the brain functions that regulate speech production. Although the exact cause is unknown, we know that if your parents stutter, you stand a higher chance of stuttering too.

Oddly enough, there are three situations in which stutterers find they can talk freely.

Stutterers with pets not only get to enjoy the companionship of their furry friends but also their therapeutic benefits. Stutterers find any anxiety about speaking evaporates when talking to their pets – presumably because they don't judge them or ridicule them. Although some cats may seem like they are …

When we sing, our words tend to be smooth and prolonged. All of us are less likely to stumble over the lyrics, which possibly explains a stutterer's ability to sing fluently.

A stutterer's ability to read steadily aloud when alone seems to come down to them feeling more relaxed in their solitude. In fact, relaxation appears to be the common element for fluent singing, reading and talking to pets.

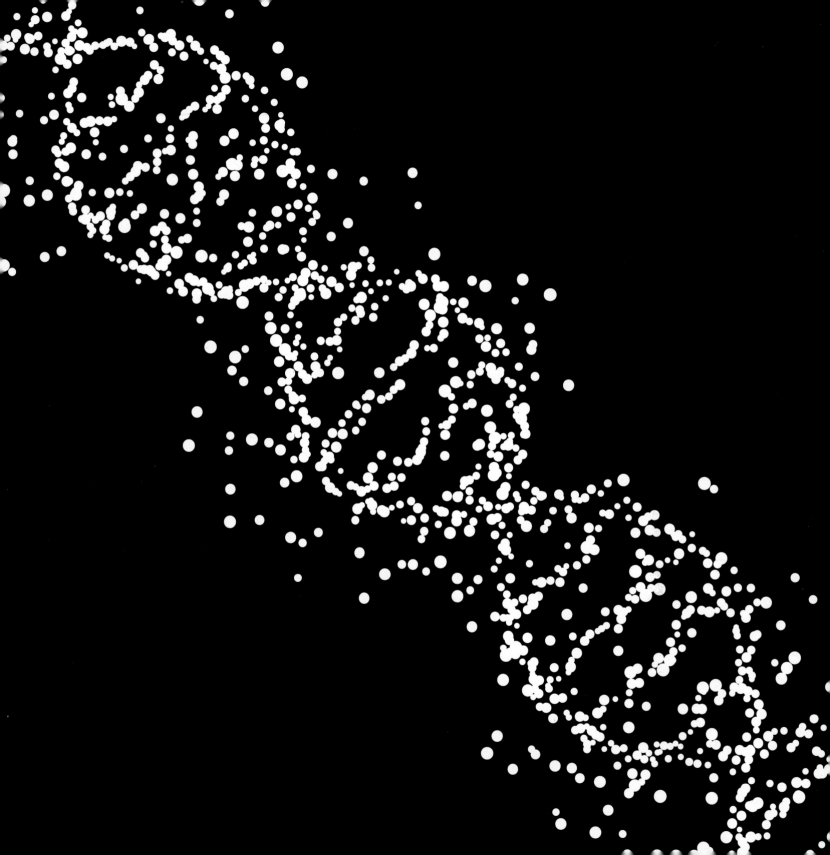

If you laid all your DNA molecules end to end, they'd span the diameter of the solar system. Twice.

Deoxyribonucleic acid – what a mouthful; no wonder we go with DNA – is the building block of life. It's made up of two long molecules arranged in a spiral, like a twisted ladder. This is commonly referred to as the double-helix structure. The blueprint for all life, DNA is present in the nucleus of every single cell in your body. It carries your genetic information and has the instructions that a living organism needs to grow, reproduce and function. Don't leave home without it.

The DNA inside each of your cells is compressed into a space much smaller than the eye can see. Arranged in a tightly packed coil are three billion base pairs, and they fit into a space only six microns across. That's incredibly small when you realise that the average cross-section of a human hair is around 50 microns.

Yet if you took a single cell of DNA and stretched it out, you'd have one extremely thin two-metre-long thread of human. We have around ten trillion cells in our bodies, and the total length of all this DNA laid end to end would be a whopping 20 billion kilometres.

When it comes to measuring the diameter of the solar system, there's some debate about where the boundary actually is. If you go with the edge of Pluto – as is generally accepted – the diameter is around ten billion kilometres.

That's handy: it's just long enough for us to loop our DNA strand around it twice.

There's a volcano on Mars that's three times as tall as Mount Everest.

Around 4000 brave souls are known to have reached the summit of Mount Everest. When they describe the experience, these individuals commonly say it was like standing on top of the world. Given that Mount Everest is the tallest mountain on our planet, they have! Located in the majestic Himalayas, Mount Everest rises almost 9 kilometres above sea level, and its summit spans the border between Nepal and China. This mountain is so big, it needs two countries to live in.

By the way, there are actually taller mountains on Earth – like the Mauna Kea volcano in Hawaii. But since a large portion of this lies under the ocean, Mount Everest takes the title of the tallest landmass above the sea.

Amazingly, our red neighbour, Mars, has a volcano three times taller than Mount Everest! What a show-off …

Olympus Mons is the almost unimaginably gigantic Martian volcano. It's bigger than the US state of Arizona and almost as big as France! This volcano is so huge that it sticks up out of Mars's atmosphere.

This massive formation is a shield volcano, meaning that it was created by lava slowly flowing down its sides. Because of this, Olympus Mons has a low, squat appearance, curving with the planet. If you were to climb it, you'd feel like you were walking normally the whole time. The same couldn't be said if you were climbing Mount Everest, where it would very much feel like an ascent.

Its gargantuan size can also be blamed on a combination of Mars's lower surface gravity and the high eruption rates. It's a fairly young volcano too and, as such, may still be active, meaning potentially more Martian fireworks ahead.

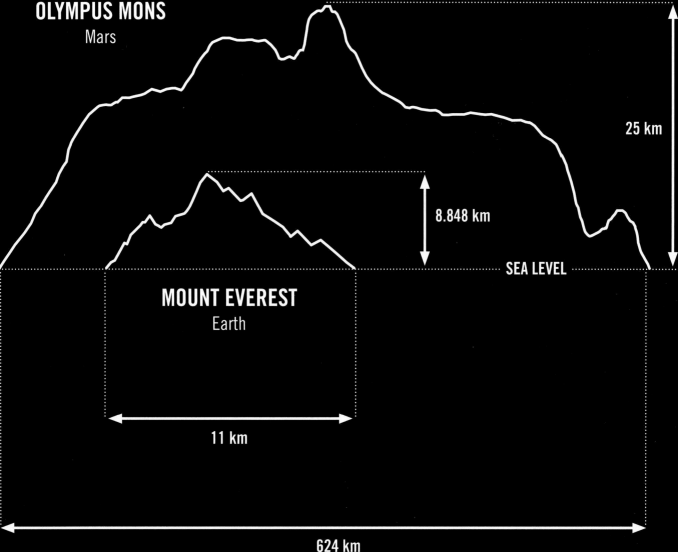

OLYMPUS MONS
Mars

25 km

8.848 km

SEA LEVEL

MOUNT EVEREST
Earth

11 km

624 km

The last execution by guillotine happened the same year Star Wars came out.

Having your head chopped from your body by a giant falling blade sounds like a horrific way to die. Execution by guillotine is a barbaric type of execution we usually associate with a long distant past.

It's pretty crazy to think that the guillotine, or 'National Razor' as it was often known, remained France's standard method of judicial execution until the abolition of capital punishment in 1981. Named after Dr Joseph-Ignace Guillotin in 1789, the guillotine was a response to the cruelty of botched sword and axe beheadings.

Guillotines also provided a class-system in execution methods. While nobles were decapitated, common-folk were a little less fortunate, and were often burned to death or broken on a wheel.

This wheel involved breaking a criminal's bones with a giant wheel and/or bludgeoning them to death with it. Painful doesn't even begin to describe it …

So many people were executed by guillotine in France that the time period was known as the 'Reign of Terror.' Not even the royals Marie Antoinette or her husband King Louis XVI were spared.

The last person to be executed by guillotine was Hamida Djandoubi in 1977. This also happened to be the very same year that George Lucas's epic space opera *Star Wars* was first released, very quickly becoming a worldwide pop-culture phenomenon.

Decapitated heads, and laser fights in space. 1977 was quite the year.

Earth's largest land organism is a 2400-year-old fungus.

Most tourists wandering through Malheur National Forest in the US state of Oregon are blissfully unaware that below them lives one of the world's living wonders. And it's a fungus. Not just any fungus, but a parasitic fungus. *Armillaria ostoyae* covers an area of more than 8.8 square kilometres. Plus it's still growing. And at well over 2400 years old, it's ancient too!

The Humongous Fungus generates honey mushrooms, which grow above ground through autumn. But while these mushrooms show us the extent of the fungus's coverage, they're only the fruit body of a much larger underground network called the mycelium. This forms a subterranean mat of spreading threads that draw nutrients from nearby vegetation, with which they feed the fungus.

Over the course of hundreds and thousands of years, as the Humongous Fungus grew, any tree or shrub unlucky enough to be in its slow-moving path has been infected, killed, eaten and then engulfed. These mycelia don't ever stop.

Behind them they leave an ever-expanding network of dead and dying trees – a kind of road map of the Humongous Fungus's growth. In fact, this was how it was discovered in the first place. Investigating the loss of large numbers of trees, scientists took fungal samples from the park. They discovered, to their astonishment, the fungal samples all had the same DNA.

Because of low competition for land and nutrients, this organism had kept growing, covering an increasingly vast geographical area. The scientists suddenly realised they were looking at the world's largest known living thing.

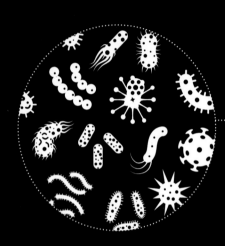

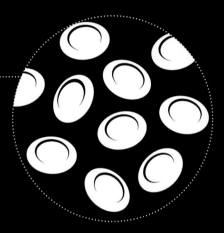

**40 TRILLION
BACTERIAL CELLS**

Dominated by colon bacteria

**30 TRILLION
HUMAN CELLS**

Dominated by red blood cells

More than half of your body isn't human.

It's pretty nuts that only 43% of our body's total cells are proper 'human cells'. The other 57% are microscopic tenants: they moved in without asking you and definitely aren't paying rent. These microscopic creatures include bacteria, archaea (similar to bacteria but with a different molecular structure), viruses and fungi, and cover almost our entire body. The biggest population lives in our bowels. We refer to this collection as our 'microbiome', and it's fundamental to our wellbeing.

Our microbiome plays a major role in our digestion, regulating our immune system, protecting us against disease and manufacturing the vitamins that are vital to our survival.

Until recent times, humans viewed the microbial world as something that worked against us. We now understand that these microbes transform and benefit our health in numerous different ways.

Of course, if these microscopic colonists constitute a greater proportion of our cells than our own human ones, it does beg the question, 'Are we more microbial than human?'

There currently appears to be no definitive answer to this big question. But maybe the symbiosis between us and our freeloading bacteria holds the key. They've adapted to changes in our diets, health and lifestyle choices, and our bodies have adapted to them. Our tenants, it seems, have helped shaped the very path of human evolution.

Pluto hasn't completed a full orbit since being discovered in 1930.

Relative to Earth, Pluto is a cosmic slow coach. This dwarf planet takes 248 Earth years to complete a single journey around the Sun!

It's likely that Pluto has existed since the beginning of the solar system, making it about 4.5 billion years old. It's reasonable to assume that Pluto has already completed a few million trips around the Sun — just very, very slowly. However, since being discovered by astronomer Clyde Tombaugh in 1930, Pluto hasn't even completed a single revolution!

In the time that we've been friends with Pluto, we've only seen it crawl through a third of its orbit. It will be 2178 before Pluto is back to where it was in its orbit when Tombaugh spotted it in 1930. That's an awfully long time between drinks.

Pluto isn't the only planet on a go-slow. Neptune, the eighth and farthest planet from the Sun has managed to complete only one full orbit since its discovery in 1846. To be exact, Neptune takes 165 years to orbit the Sun. No wonder these cosmic neighbours hang out!

One reason Pluto's orbit is so slow is that it goes the long way. Compared to the other planets, its orbit is more oval-shaped, or elliptical. That means that the distance between Pluto and the Sun changes; occasionally it is closer to the Sun than Neptune is.

But no matter how you look at it, Pluto is a laggard. The two closest planets to the Sun — Mercury and Venus — leave it in their dust as they complete their orbits of the Sun in 88 days and 225 days respectively.

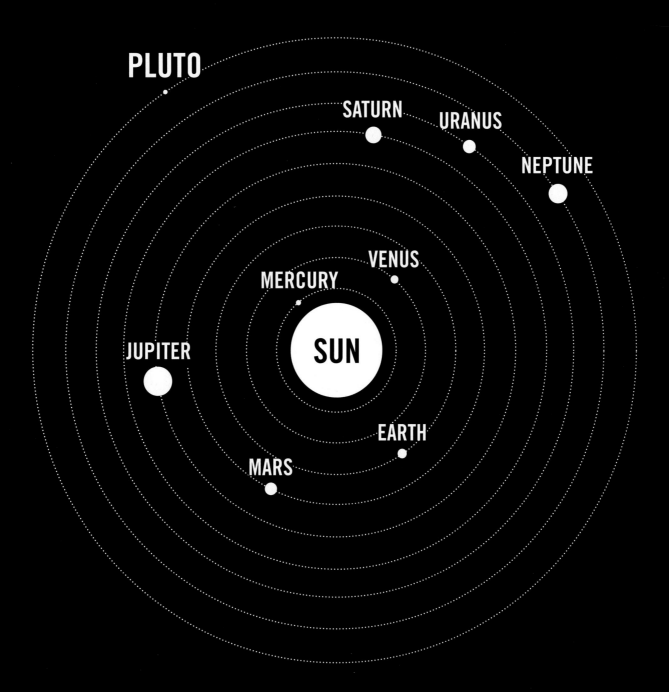

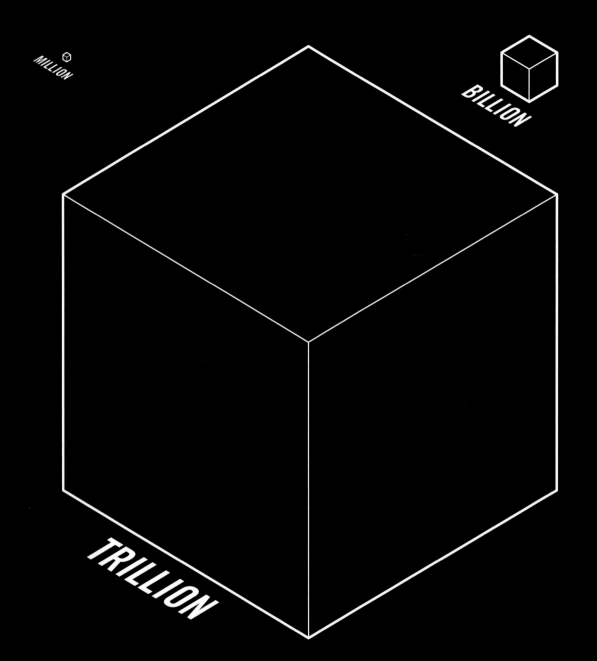

MILLION

BILLION

TRILLION

A million, a billion and a trillion aren't remotely similar in size.

People tend to group big numbers together. Maybe it's because they look similar. Maybe it's because they rhyme. After all, who doesn't love a good rhyme. Thinking that big numbers come in groups could not be further from the truth, especially in the case of million, billion and trillion. And in case you're unsure, in Australia a billion is a thousand millions (1,000,000,000), and a trillion is a thousand billions (1,000,000,000,000).

Look at it this way: if you were able to earn a dollar every second you were alive, you'd be an infant millionaire in little under two weeks, complete with gold-lined nappies. After hitting your precocious millionaire peak, you'd be waiting until you were 31 years old to make billionaire status. But, sadly, those dreams of becoming a trillionaire are never going to happen. It would take a shocking 32,000 years to achieve this level of wealth. That's both mind-blowing *and* depressing …

Here's another way of looking at it. If you took your infant wealth and changed your theoretical millions into crisp one-dollar bills, your million-dollar stack would be the height of a chair. It's not surprising that your billion-dollar pile would be much more impressive, soaring a kilometre into the sky — even higher than Burj Khalifa, the world's current largest building!

But even more impressive would be your unachievable trillion-dollar pile. As this builds up, it would leave its Earthly confines, stretching up 1000 kilometres — way past the Kármán line, the altitude where space begins. It would briefly draw level with the International Space Station and keep going until it was two-and-a-half-times that distance from Earth!

Nature is teeming with the world's strangest and most powerful hallucinogenic molecule.

You really cannot imagine a weirder molecule than N,N-Dimethyltryptamine (DMT) or a stranger experience when you take it as a drug. DMT is a psychedelic tryptamine compound. When inhaled or taken intravenously, DMT gives the user a completely immersive and out-of-body experience that can feel more real than normal consciousness. This high often includes complex, vivid, fractal-like hallucinations that defy description, mainly because we simply don't have the language to articulate them.

Not only does DMT cause mind-blowing hallucinations, but a majority of users have reported meeting — while on the drug — intelligent beings. These entities, or 'self-transforming machine elves', have been described to wordlessly convey messages of love or present wonderful, intricate inventions resembling cosmic Fabergé eggs.

All this sounds pretty wild, but what's even wilder is that DMT is absolutely everywhere. Including inside you. Mysteriously, our bodies produce the chemical, and it also occurs naturally in the plant and animal kingdoms — from mammals to marine animals, grasses to bark, toads to frogs, mushrooms and moulds to flowers and roots.

It's illegal to possess DMT, which is rather strange as it means that everyone is technically breaking the law.

Currently, government-sanctioned global research into DMT is taking place at medical establishments such as Johns Hopkins School of Medicine, Baltimore, and Imperial College London. The short-acting psychedelic offers a promising new treatment pathway for individuals suffering depression and anxiety.

The galaxies we see from Earth are millions of years old, and many are already dead.

If you look up on a moonless and star-filled night, the mind-blowing sight of thousands upon thousands of twinkling dots will illuminate the darkness. But did you know that as you gaze into the night sky, you're doing a Marty McFly–*Back to the Future* and looking back in time? And it's all thanks to the finite speed of light.

When we observe objects extremely far away from us, the light that hits us has started shining a long time ago — it could have been travelling through space for billions of years. Essentially, this means we aren't actually seeing how the object appears now, but how it appeared when the light was originally emitted.

Aside from our own Milky Way, Andromeda is the only other galaxy that we can see from Earth with the naked eye. It's our closest galactic neighbour at just 2.5 million light years away — that's practically next door in cosmic terms! When the light from Andromeda finally reaches Earth, we're seeing an image of the galaxy that's 2.5 million years out of date. For all we know, any of the galaxies and stars we're able to observe are long gone.

Like everything else that exists in the universe, the Andromeda and the Milky Way galaxies are attracted to one another by gravity. The consequence is that in four billion years' time, they'll collide, merging together to form an even bigger galaxy!

Formula 1 drivers lose up to five kilograms per race.

The sport of Formula 1 racing is all about extremes. The drivers' world features extreme speeds, extreme danger, extreme travel, extreme egos, extreme money and, more surprisingly, extreme weight loss. A driver can lose five kilograms during a two-hour race! This is all water lost through sweating, thanks to extreme heat. On average, temperatures in the cockpit reach a sweltering 50°C!

It doesn't help that F1 drivers have two layers of clothing: they wear a fire-retardant suit under their racing suit. But safety comes before comfort.

Such severe water loss can impact a driver's physical and psychological abilities – not ideal when hurtling around a racetrack at speeds of up to 350 km/hour. This is why drivers attempt to keep their fluid levels in check. They drink a lot of water before and during every race.

F1 car cockpits have water bottles with a pipe that goes directly through the driver's helmet, so they can slowly sip water with mineral salts to combat the inevitable dehydration.

As well as the heat, drivers experience high g-forces during a race. They can hit 5gs while braking, 2gs while accelerating, and between 4gs and 6gs while cornering. This would be like 30 kilograms of force pulling their head to the side.

The deceleration experienced by an F1 driver is the equivalent of driving a car straight through a brick wall. Ouch.

The strain a driver's body is subjected to only increases their energy consumption and adds to the staggering weight loss.

STEGOSAURUS

LATE JURASSIC PERIOD
144 – 156 million years ago

83 MILLION
YEARS

TYRANNOSAURUS REX

LATE CRETACEOUS PERIOD
65 – 67 million years ago

67 MILLION
YEARS

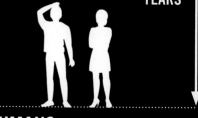

TODAY

HUMANS

The Tyrannosaurus rex lived closer to our time than it did to the Stegosaurus.

The dinosaurs not only existed a long time ago, but they also existed for a very long period of the history of our planet. They ruled the Earth for almost 175 million years — that's a pretty good run in anybody's book. The reign of the dinosaurs is even more striking when you consider that modern have only been around for 200,000 years!

So far over 700 different species of dinosaur have been identified and named. However, palaeontologists believe that there are many more new and different dinosaur species yet to be discovered.

Two of the best known species are the *Stegosaurus* and the *Tyrannosaurus rex*, and it's easy to picture these dinosaurs battling furiously for supremacy. Although, thinking these famous dinosaurs were around in the same time period would be a mistake.

In truth, they never even had the pleasure of meeting. Which was probably a good thing for the plant-eating *Stegosaurus*.

The *Stegosaurus* roamed the Earth during the late Jurassic period — between 144 and 156 million years ago. The *Tyrannosaurus rex*, on the other hand, lived during the late Cretaceous period — about 65 and 67 million years ago.

What this means, startlingly, is that the period in which the *Tyrannosaurus rex* lived is closer to humanity than to that of its fellow dinosaur. In fact, to a *Tyrannosaurus rex*, a *Stegosaurus* would appear positively ancient. Though every bit as delicious, we can suppose.

Not even light can escape a black hole.

Why was the black hole hungry?
Because it had a light breakfast!

Black holes occupy a point of space where gravity is so strong that absolutely nothing can escape its pull. Not particles. Not electromagnetic radiation. Not even light.

Given that light has no mass at all, the gravitational pull of a black hole must be incredibly strong.

Because no light can escape black holes, they are extremely difficult to see. We only know they exist from watching what happens to the unfortunate stars that stray too close. The event horizon of a black hole is where gravity becomes too powerful and the star's matter is pulled in.

These cosmic monsters form when the centre of a massive star collapses in upon itself, crushed under the weight of its own gravity. This collapse triggers an exploding supernova that blasts part of the star deep into the vastness of space.

These are called stellar-mass black holes, and they can be up to fifteen times the size of our Sun! Then again, this is small change compared to supermassive black holes, which are often hundreds of thousands to billions of times bigger. We're unsure of their origins, but we do know that they exist in the centre of nearly all galaxies.

The supermassive black hole in our own Milky Way Galaxy is a stunning four million times the mass of our Sun!

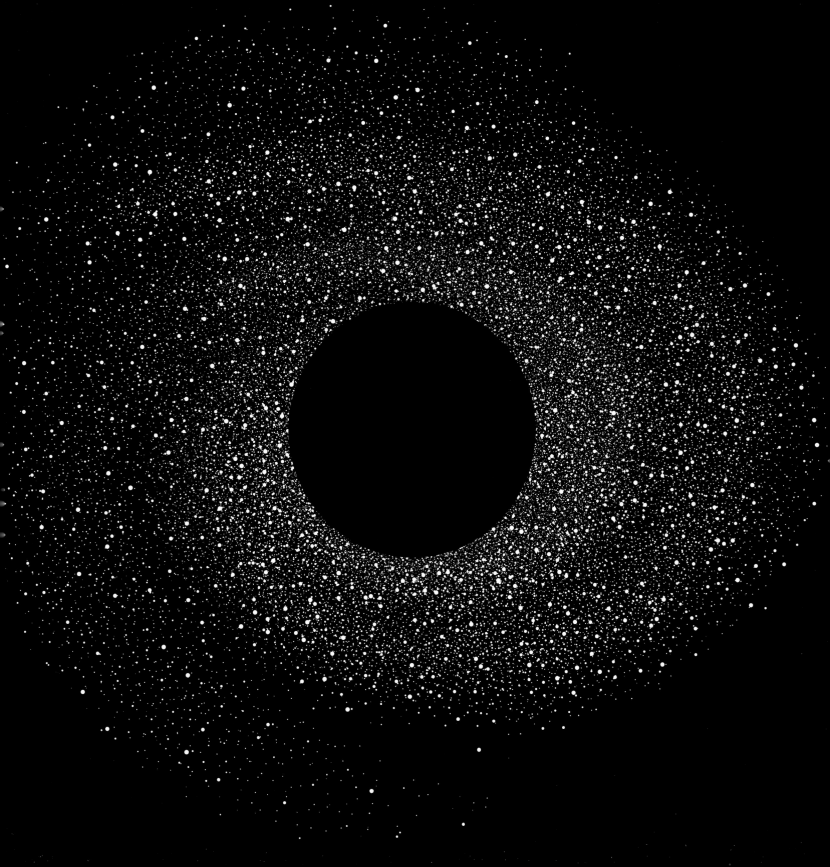

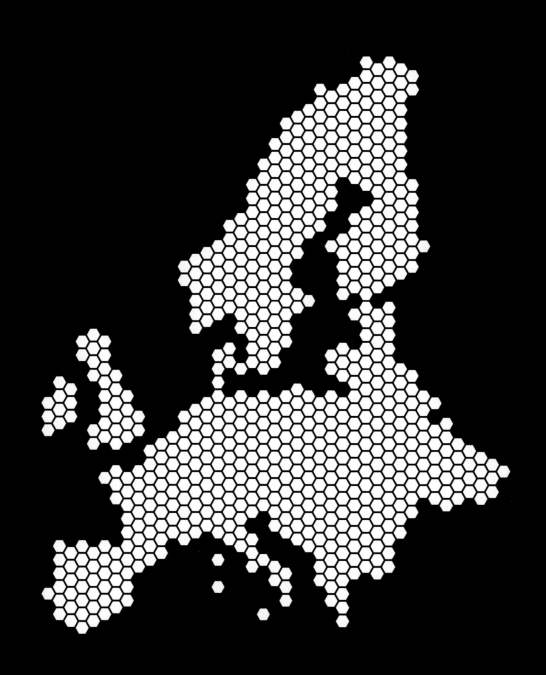

All Europeans today are related.

Calling the people of Europe one big family seems like a hard claim to make, given their centuries of fighting and constant disagreements about whose cuisine is best. But that's exactly what genetic sequencing has shown us. Europeans are basically one big brood, closely related to one another. There's common genetic material between countries as far apart as Britain and Greece, after the great population movements of the first millennium, like the Saxon and Viking invasions of Britain.

All Europeans alive right now are related to the same set of ancestors from as little as 1000 years ago. In other words, if you were alive 1000 years ago and left any descendants, then you'd be a distant ancestor of every European today.

Since the number of ancestors of each European has roughly doubled with every generation, you don't need to go too far back to find someone who featured in all of their family trees.

And what's even more remarkable is that you wouldn't have to go back much further than that to a point where everyone in the world is related to one another – possibly as little as 3000 years ago! Practically yesterday in evolutionary terms.

It's heart-warming to know you don't have to look back in time too far to discover we're all one big family – even if it is a dysfunctional one …

There are only 66 years between the first plane flight and man landing on the Moon.

It may have only lasted for 12 seconds and covered a distance of 37 metres, but the airborne jaunt that Orville and Wilbur Wright took in 1903 made them the first humans to ever complete a powered, sustained and controlled airplane flight. We had lift-off!

The two brothers who initiated the era of human air travel were former bicycle technicians from Ohio, US. Their first aircraft, the *Wright Flyer*, was born in North Carolina and built with just wood and fabric!

As was proper in those times, both brothers wore dark suits, white shirts with a necktie and a cap for their history-making foray into the skies. No singlets and boardshorts allowed.

In 1969, only 66 years later, Neil Armstrong was dressed a little differently as he piloted fellow astronauts Edwin 'Buzz' Aldrin and Michael Collins in the Apollo 11 spacecraft.

Practically the whole of humanity followed their progress, gathering round radios and televisions as Armstrong became the first human to ever set foot on our orbiting celestial neighbour, quickly followed by Aldrin.

It was an awe-inspiring feat, one that still resonates today. Only eight years earlier, US President John F. Kennedy had declared that the United States should commit to the goal of landing a man on the Moon and returning him safely to the Earth.

The Apollo 11 team spent three days travelling through space to reach their destination, and almost the same amount of time to get home. A lot longer than the Wright brothers took with their equally historic flight.

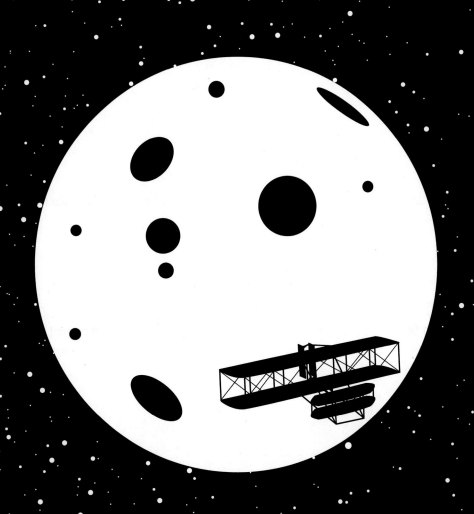

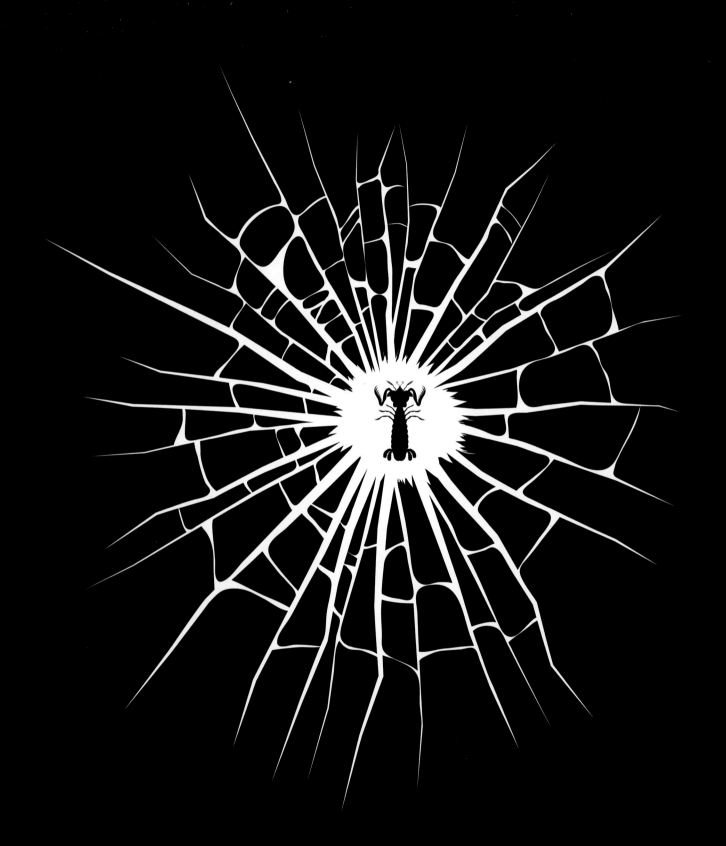

The mantis shrimp can throw the fastest and most powerful punch in the entire world.

If you're unlucky enough to count yourself an enemy of the mantis shrimp, then you'd better learn how to take a punch, and fast. Although, to be honest, you'd be wasting your time — you wouldn't even see that right hook coming … These shrimp may be small, but they can punch at 80 km/hour, landing that blow with the same force as a rifle bullet. This punch would definitely leave some imaginary little birds chirping around your head.

It's even more mind-blowing when you realise this stunning punch is thrown in water, moving rapidly through considerable drag and resistance. Nothing can get in the mantis shrimp's way as it delivers its thunderous wallop in three thousandths of a second.

The shrimp's power and speed come from the large muscles in its upper arm, which fire like a spring, accelerating at up to 10,000 times the force of gravity! So quickly does that arm move through the water that it lowers the pressure, pushes the water to boiling point and produces flashes of light. Further damage is inflicted as the water pressure returns to normal and a tremendous amount of energy is released.

These tough-guy crustaceans use their power to feast on crabs, snails and small fish but are known to take on large fish and even the occasional octopus! They also have a reputation for cracking aquarium glass and damaging boats.

The mantis shrimp is one little fighter you do not want to mess with.

Oxford University is older than Aztec civilisation.

What we know of the ancient Aztec civilisation of Central Mexico is fascinating. The discovery of its ruins, deep in the jungle, by Indiana Jones-type archaeologists adds to its mystique. Visitors to museums displaying Aztec exhibits are awestruck by the relics of the Aztecs' past, which include intricate calendar stones and representations of the feathered serpent, Quetzalcoatl.

The city of Tenochtitlán was founded by the Mexica people at Lake Texcoco in 1325. Tenochtitlán was captured by Spanish conquerors in 1521, a mere 196 years later.

While this civilisation is most definitely old, it hasn't got as many candles on its birthday cake as one of the oldest learning institutions in the world – Oxford University in England.

Take a walk around Oxford University and the signs of its longevity are everywhere. As the first university in the English-speaking world, Oxford is a unique and venerable institution.

There's no clear date of foundation, but some form of teaching existed at Oxford from as early as 1096. Some of its current professors even look like they were teaching back then, too …

By 1249, Oxford had grown into a fully fledged university, complete with student housing. Its three original halls of residence were University, Balliol and Merton colleges.

Today it has produced 58 Nobel Prize winners and 27 prime ministers, and is often cited as the best university on the planet.

SUN

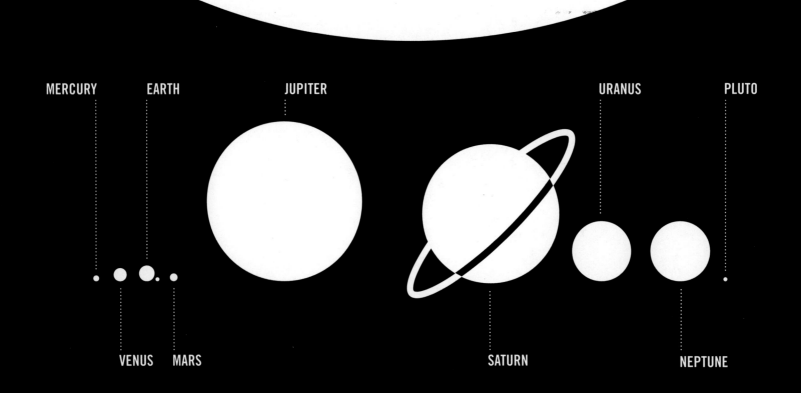

MERCURY   EARTH   JUPITER   URANUS   PLUTO

VENUS   MARS   SATURN   NEPTUNE

The Earth fits into the Sun more than a million times.

All life on Earth depends on light and energy from the Sun – that big ball of gas at the centre of our solar system. It's no wonder the ancient Egyptians worshiped it as a god ...

The Sun is heated by nuclear fusion. Inside its core, hydrogen nuclei continuously melt together to form helium nuclei, releasing vast quantities of energy. Away from the core, this energy is converted to light, which can take up to 100,000 years to escape. Weirdly enough, once that light has escaped the core, it takes only eight minutes to travel the 150 million kilometres to reach us!

Thinking about how powerful the Sun's effects are here on Earth probably gives you an indication of its size. It's by far the largest object in our solar system, making up a ridiculous 99.8% of our total mass.

In fact, the Sun is so big you could fit more than a million Earths inside of it – 1.3 million to be exact!

The surface of the Sun is marked with small black dots, called sunspots, that are slightly cooler than the mind-blowingly hot 5500°C temperature of the overall surface. Sunspots range in size from 1500 to 50,000 kilometres in diameter – bigger than Earth!

Don't forget, however, that the Sun is classified as a yellow dwarf star. As the name suggests, there are huger stars out there in the universe.

If our Sun is a god, surely it's an angry one. In around five billion years, this yellow dwarf will expand into a red giant, 150 times its present size. As it swells, it will vaporise and destroy the Earth and everything else in its vicinity.

Every single dog breed today shares a common ancestor with the grey wolf.

You only have to spend five minutes at a dog show to realise that dogs come in the most spectacular array of shapes, sizes, colours and coats. In fact, there are over 340 different breeds of dog today! It's pretty extraordinary that all these different breeds share a single common ancestor with *Canis lupus* – the grey wolf.

Scientists believe this common ancestor was a prehistoric wolf that lived in Europe or Asia as long as 34,000 years ago. Dog skulls dating back 33,000 years have been found in caves inhabited at the same time by humans. This indicates that the dog was the first species to be domesticated, and had evolved even before humans invented agriculture!

The reasons these pooches became domesticated is still unclear. One hypothesis is that the dog's ancestors first befriended human hunter-gatherers by moving near the outskirts of their camps, scavenging for leftover food. The tamer and less aggressive wolves would be the most successful at this. Smart cookies.

It makes sense that humans would then have developed a symbiotic relationship with our furry friends and in return used them for hunting and scaring away other animals. Over time, this relationship would see them evolve into our modern-day animal besties.

In fact, it's our selective breeding of dogs that has contributed to the vast variety of breeds today. We have directly affected their evolution.

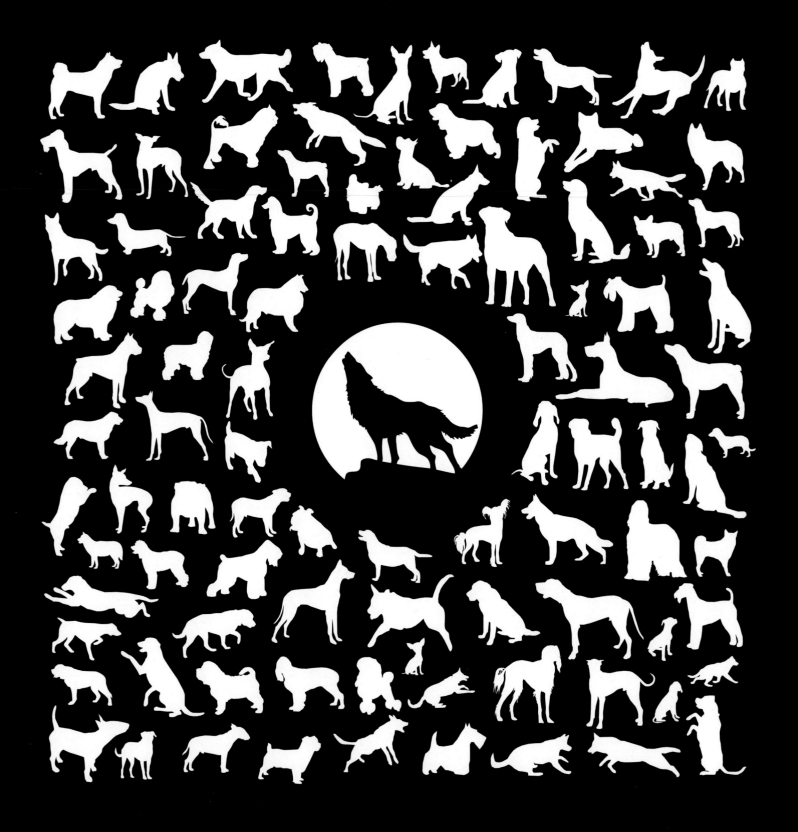

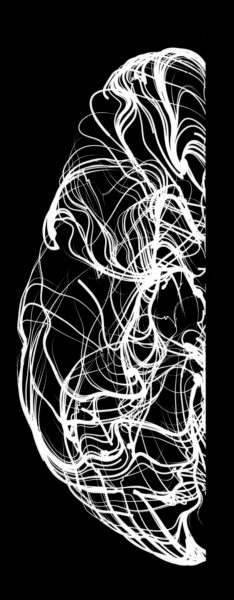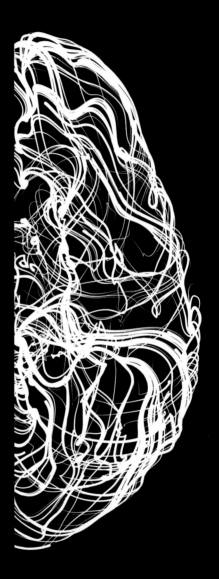

Half our brain could be surgically removed with no adverse effects on our personality, memory or humour.

The human brain is often described as the most complex thing in the universe. It's the command centre for our bodies, telling us how to move, regulating our heartbeat and breathing, and sending us to sleep. It also stores our memories, handles our emotions, forms our personalities and allows us to have independent thoughts. It's our window and filter to receive and interpret information from the world around us. The brain is pretty awesome!

All these amazing actions are carried out through an intricate network of billions of nerve cells, called neurons, which transmit millions of electrical signals.

We know more about the brain now than we ever have, but it remains one of our greatest mysteries. No-one knows, for instance, why we're able to remove half of it and carry on without adverse effects.

Hemispherectomy — the operation to remove, disconnect or disable half a brain — is a legitimate medical procedure. Astonishingly, this invasive surgery can be performed with no significant long-term effects on memory, personality or humour!

Hemispherectomies are used to treat a variety of seizure disorders localised to a single hemisphere of the brain. It's reserved for the most extreme cases — people whose severe and debilitating seizures don't respond to medications or less invasive surgeries. The procedure has a history of successfully curing seizures in between 85% and 90% of patients.

Reassuringly, hemispherectomies have evolved over the years. Sufferers no longer have to sacrifice half their brain. Thanks to modern medical developments, seizure relief can be achieved with minimal brain tissue removal. Phew!

Two particles on opposite sides of the universe can affect each other instantly.

Quantum mechanics is the head-melting branch of physics concerned with the small. The sort of small requiring the latest transmission electron microscope and matching camera. It's a body of scientific laws that describes the incredibly strange behaviour of atoms, photons, electrons and the other particles that make up our universe.

Unlike classical mechanics – which describes how things move at standard speeds and sizes, like how a tennis ball travels through the air – in quantum mechanics, objects exist in a fog of possibilities. There's a probability of being in one place and another chance of being at another. Absolutely mind-blowing.

Take the phenomenon known as quantum entanglement, for instance. When a laser beam is fired through a crystal, it causes an individual photon to be split into a pair of entangled photons. Amazingly, these two particles somehow remain linked to one another, so any action performed on one, like measuring it, will affect the other. Brace yourself, though: this reaction still happens no matter how far apart the particles are, even if they're light years across the universe! This affected change seems to take place at a speed faster than the speed of light, possibly even immediately!

The speed of light is, of course, the speed limit of the universe. The speed of change for entangled particles should not be possible, which only adds to the mystery of quantum entanglement!

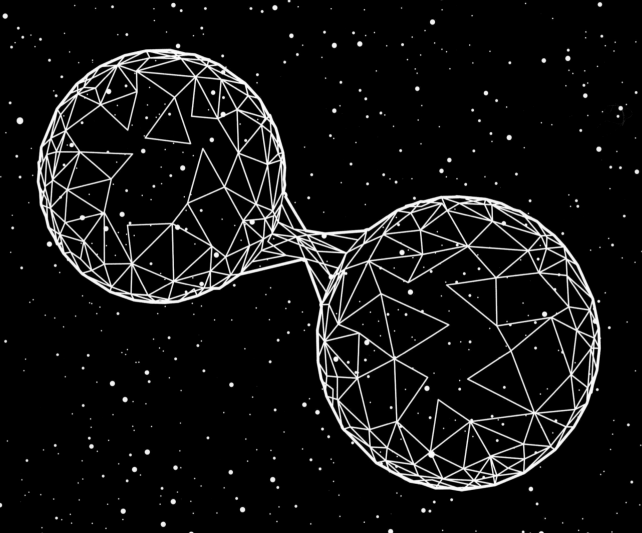

There exists a jellyfish that's biologically immortal.

Everything that lives will one day die. This is the law of nature in our universe. From the smallest creatures on Earth to the greatest stars in the cosmos, death is a certainty for all things that live.

One species has chosen to ignore the rules, however, and happily exists forever: *Turritopsis dohrnii*, the 4.5-mm transparent jellyfish. In addition to being immortal, it seems to be everywhere in our oceans.

Their secret to cheating death is reproduction. Once they procreate, these tiny jellyfish revert to their juvenile, sexually immature state. In other words, they get to experience their awkward aquatic adolescence all over again, but as teenagers who actually know a thing or two this time. Instead of locking themselves in their bedrooms and listening to loud music, these jellyfish fall to the ocean floor, where their bodies shrink and their tentacles retract.

For *Turritopsis dohrnii*, this is more than merely a phase. Their reproduction cycle can be repeated over and over again, making them biologically immortal. Provided they avoid predators and disease, they can continue this cycle forever.

These small blobs of jelly have wanderlust too. Marine biologists suspect that the immortal jellyfish originated in the Pacific Ocean, but regardless of where they started, they can now be found in the temperate-to-tropical regions of all the world's oceans. It's thought that, because of their size, these jellyfish can inadvertently hitch free rides on the ballast tanks of ships, which explains why they're slowly spreading throughout the seas. By travelling the Earth, they increase their chances of meeting other biologically immortal jellyfish and making more tiny jellies.

It's more likely we're living inside a computer simulation than not.

Simulation theory is the bizarre idea that we're all digital beings living in a vast computer simulation created by our own technologically advanced far-future descendants.

Although this sounds like the bonkers plot from a sci-fi film, it's something many scientists take seriously. So seriously that some of them say it's an inevitable truth. Blimey!

There are two main parts to the argument. The first requires a huge leap of the imagination – to accept that, at some point in the future, we'll be able to simulate human consciousness inside a computer.

The second is that any future advanced civilisation would have monumental amounts of computing power at their disposal, dwarfing anything we have today.

If they were to then use this advanced processing power to run ancestral simulations, the future civilisation would soon be far outnumbered by simulated digital counterparts. And if there are more simulated minds than organic ones, then the probability of us being the real minds doesn't look too likely.

Even if there's some truth to the simulation theory, it leaves lots of questions unanswered. Where did the original, non-simulated world come from? Why have our technologically advanced far-future descendants left so much pain and suffering in the simulation?

And then there's the huge question of what would happen if we invented an ancestral simulation that nests inside our own simulation? Or has this already happened many times over?

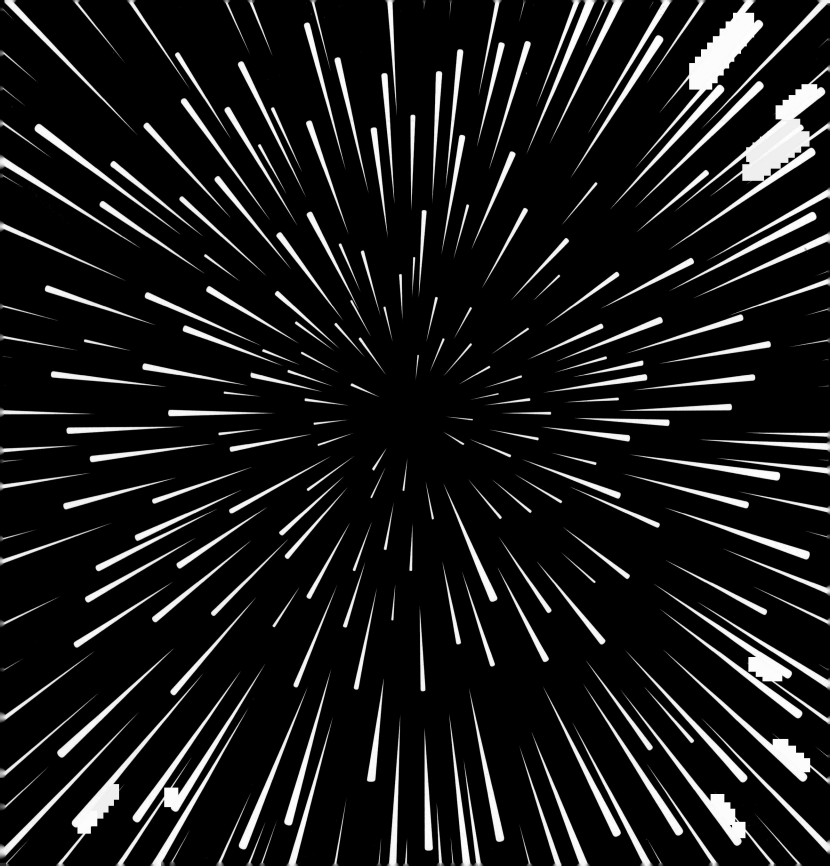

We are literally made up of stardust.

In order for life to exist on Earth, there has to be carbon, hydrogen, nitrogen, oxygen, phosphorus and sulphur. These elements are the building blocks of life, and together they make up 97% of the atoms in all our bodies.

At the centre of the Milky Way, there's an abundance of those six elements. This means that the hundreds of thousands of stars in our galaxy all share the same major elements as us – when someone next calls you a star, they probably have no idea just how true that is!

Inside the heart of stars, a process takes place called nucleosynthesis, which is basically the making of new, heavier elements through nuclear fusion. When stars run out of fuel and die – after billions of years – they collapse under the weight of their own gravity, exploding in a violent supernova and ejecting their trillions of tonnes of atoms into the vastness of space.

They blow up with such ferocity that enormous clouds of gas and dust travel from one galaxy to another, covering hundreds of kilometres in a single second. When these clouds of dust and gas fall into the gravitational pull of neighbouring galaxies, a fresh generation of stars is born. These new stars then repeat the life cycle, and more elements are created before being flung out into the cosmos when the star dies.

Stars make up every element, and if you combine those elements in different ways, you can make gases, minerals and even large chunks of rock, like asteroids. From asteroids, planets are born, and if you add a little water and the other fundamental ingredients required for life, you'll eventually arrive at us. And who knows what next!

All of us are indeed made of stardust.

Starlings fly around in swooping, intricately coordinated patterns.

Despite there being 310 million starlings in the world few people have heard of, let alone seen, them perform their mass aerial stunt known as 'murmuration'. If you had, your jaw would have hit the floor.

This phenomenon takes place at dusk, when the tiny birds form a huge, noisy, swooping flock then proceed to trace a series of intricately coordinated patterns in the sky.

Murmuration could be mistaken for some kind of avian performance art but, in reality, it has more to do with basic survival than it does with dramatics. By teaming up in such massive, fast-moving groups, the starlings enjoy safety benefits.

Unsurprisingly, predators find it almost impossible to target one starling in the middle of a hypnotising flock.

Incredibly, starlings in a group are aware that those on the outside and those that land first are the most vulnerable to attack. Their response is to create an internal competition to weed out the weakest of the flock.

While murmurations are often made up of thousands of starlings, there have been reports of flocks that number in the hundreds of thousands. That would be a truly mesmerising sight!

In 1977, Earth received a mysterious radio signal from outer space.

In 1977, astronomer Jerry Ehman was scanning the sky for possible signs of extraterrestrial life and, to his amazement, he found something. While pointing the Ohio State University's Big Ear radio telescope to a cluster of stars called Chi Sagittarii, he received a strong 72-second blast of radio waves. Ehman was so astonished, he wrote 'Wow!' in red pen on the readout.

This became known as the 'Wow! signal' and it still baffles us as much today as it did more than forty years ago. We immediately knew that it had come from interstellar space, but that was all we could tell from the signal.

The signal is generally attributed to extra-terrestrial intelligence, but oddly enough has never been observed since, making it a mind-blowing cosmic mystery. A recent theory about the signal's origin was that it was generated by two comets discovered only ten years ago. Under certain conditions, comets will emit radio waves from the gases surrounding them as they zoom closer to the Sun.

While these two comets were in the right place at the right time in 1977, there is still doubt about this explanation. Sceptics claim that the type of signal received would be entirely different had it been made by comets. And so the mystery remains.

Cleopatra's time was closer to the iPhone era than to the building of Giza's Pyramid.

It's usual to associate the legendary Egyptian queen Cleopatra VII – renowned for her love affairs with Julius Caesar and Mark Antony – with the Pyramids. But maybe we've been jumping to conclusions.

Check the dates and you'll see that the Great Pyramid of Giza was constructed over a ten to twenty-year period, concluding around 2560 BC. Cleopatra, the last true pharaoh of Egypt, ruled from 51 BC to 30 BC. In other words, her reign took place about 2500 years after the Great Pyramid was built!

If you travel in the opposite direction along history's timeline, a significant milestone occurred in 2007: Apple's now ubiquitous iPhone first hit the shelves. It has since gone a long way to making Apple one of the world's most valuable publicly traded companies. At the time of writing, it has sold more than 1.2 billion units.

Here's where things start to get really interesting, though. When you do the numbers, the length of time between Cleopatra and the creation of the iPhone is 2037 years. As improbable as it may sound, Cleopatra lived closer to the creation of the iPhone than she did to the building of the Great Pyramid of Giza.

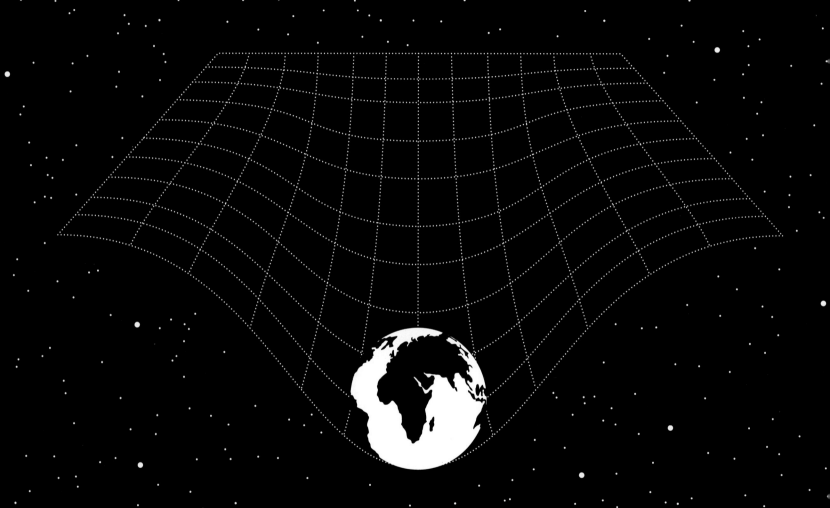

Heavy objects can distort the fabric of space and time.

It was the legendary Sir Isaac Newton who discovered that gravity is a force of attraction that pulls together all things in the universe. Gravitational pull depends on how large and close one object is to another. For example, the Sun has a lot more gravity than the smaller planet, Earth, but because of our proximity, instead of being pulled towards the Sun, we stay planted on the planet's surface. Thanks for that, Earth.

The genius of Newton was to tell us what gravity is. To figure out how it works required another genius, Albert Einstein. Turns out it wasn't a force, as Newton had supposed, but something far stranger than that. With his General Theory of Relativity, Einstein developed a mind-blowing new way to describe gravity and what causes it.

What Einstein discovered is that gravity is a natural consequence of a mass's distortion on space and time itself. In his theory, he expressed this as 'space–time'. Any object distorts the fabric of space–time, and the bigger the object, the greater the effect. The object can warp, bend, push or pull the space–time around them.

Just as a bowling ball placed on a rubber sheet stretches the material, so too do planets and stars warp space–time. Any marble rolling along the same rubber sheet will be drawn towards the bowling ball. In the same way, planets orbiting the Sun aren't actually being attracted by it, they're following the curved space–time deformation caused by its massive weight. The speed at which they're travelling is the reason the planets don't fall into the Sun.

Shark embryos eat their siblings in the womb.

Jaws has a lot to answer for. Ever since the movie's release in 1975, sharks have had a nasty image problem. Long before Spielberg's masterful take on the horror genre, many people feared these often misunderstood predators so much that they refused to set foot in the ocean.

Yet it's hard to improve the reputation of a creature that practises siblicide. Not only do sand tiger sharks eat their brothers and sisters in the womb, but they'll devour any unfertilised eggs too!

The moment a sand tiger shark embryo appears in its mother's uterus, it enters into a mind-blowing battle to the death with its siblings. As soon as they develop their embryonic teeth, the deadly quest for survival begins. Just one sibling will emerge victorious after eating its brothers and sisters.

A shark mother has two wombs, each producing several dozens of eggs, but by the time she goes into labour, there are only two babies left to deliver – one from each uterus. And giving birth is quite a feat because the foetuses are a metre long. That's because these offspring supercharge their growth with their cannibalistic pre-birth diet: winner, winner, lots of dinner.

Once born, shark pups are totally independent. They swim away from their mothers instantly, perhaps to avoid being eaten themselves; no trust could be built in that womb. They're all alone, finding their own food and avoiding predators – including others from the same species. It really is a shark-eat-shark world.

Adolf Hitler was nominated for the 1939 Nobel Peace Prize.

The name Adolf Hitler isn't one you'd place alongside historical luminaries such as Mikhail Gorbachev, Mother Teresa, Mahatma Gandhi, Barack Obama and Al Gore. But in the list of Nobel Peace Prize nominees, that's exactly what you'll find. And it's definitely not a typo ...

In 1939, shortly before he started World War II by sending his German troops to invade Poland, the Führer of Germany was nominated for the Nobel Peace Prize. However, context is everything. Hitler's nomination came from Erik Gottfrid Christian Brandt, a Social Democratic member of the Swedish parliament. Brandt was fiercely anti-fascist, and his nomination of Hitler was intended to be ironic. He was attempting to brand him as the number one enemy of peace in the world.

That wasn't all he hoped to achieve by putting forward the führer. It was also his commentary on the nomination of Neville Chamberlain, then Prime Minister of Great Britain, whom he considered undeserving of the prize. Brandt also aimed to enrage the Nazi party and highlight the fascist sympathies in his own Swedish government at the time.

The wording of the nomination was heavily sarcastic. It repeatedly referred to Hitler's quest for 'peace on Earth' and even described Hitler's famous book *Mein Kampf* as the best and most popular piece of literature in the world, next to the Bible.

Despite Brandt's off-the-charts mockery, large numbers of people completely missed this and took the nomination literally. The upshot was that Brandt himself was accused of being a fascist!

The Gympie Gympie plant produces a sting so painful it drives people to suicide.

One of Australia's claims to fame is its many creatures that would happily end your life. Sharks, spiders, jellyfish and snakes are just a few of the deadly creatures that you definitely never want to come face to face with.

Not only are there potentially lethal animals, but Australia is also home to a terrifyingly dangerous plant. *Dendrocnide moroides*, or the Gympie Gympie plant, is found in the rainforest areas in the north-east of Australia.

At first glance, this plant appears completely harmless, but on closer inspection you'd see its covering of stinging hairs. They're nothing like your average nettle hairs, though; this is the most poisonous of the Australian species of stinging trees, and it's capable of delivering a potent neurotoxin.

When touched, these hairy leaves let the neurotoxin penetrate the skin. Immediately, an extremely painful stinging sensation follows. The sting has been described as like being burned with acid and being electrocuted. At the same time. And it can remain like this for anywhere from days to years.

And this is why *Dendrocnide moroides* is also called the Suicide Plant: the agonising sting can last for so long that some victims eventually take their own lives to avoid the excruciating pain.

If you're stung, you'll instantly vomit. Also, any attempts to remove the hairs will prove incredibly difficult as they're too fine and densely packed to pluck out with tweezers. For the sake of your sanity, avoid the Gympie Gympie at all costs!

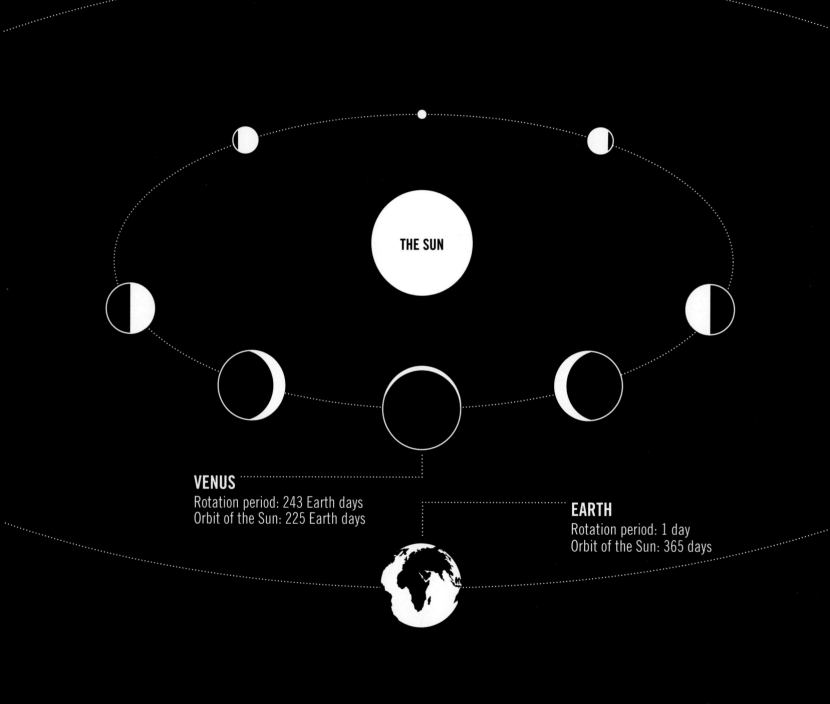

THE SUN

VENUS
Rotation period: 243 Earth days
Orbit of the Sun: 225 Earth days

EARTH
Rotation period: 1 day
Orbit of the Sun: 365 days

A day on Venus lasts longer than a year on Venus.

Venus is often called Earth's twin planet. Not only is it our cosmic neighbour but, like us, Venus is a terrestrial planet, composed mainly of metal and silicate rock. The two also orbit the habitable zone of the same star – the Sun. Commonly called 'The Goldilocks zone', it's where the temperature is just right – not too hot and not too cold – for liquid water to exist on the orbiting planet's surface. The commonalities don't end there. Both worlds are a similar size, have a similar surface composition and possess an atmosphere with a complex weather system.

However, if Venus is our twin, it's an evil one. For starters, it rotates in the opposite direction to us. That's nothing compared to Venus's atmosphere, which is over 90 times denser than that of Earth's and is primarily made of carbon dioxide and small amounts of nitrogen. This creates a pronounced greenhouse effect.

No wonder Venus is the hottest planet in the solar system. The surface temperature is more than 470°C! If we were to stand on its surface, we'd be crushed by the pressure and incinerated in seconds. How nice.

Yes, Venus is a truly nightmarish place. If you were ever to find yourself stuck there waiting for help, don't get too excited when you hear that it will be with you by the end of the day. Venus takes longer than any other planet in the solar system to complete one revolution on its axis, which is how we define a day. It rotates so slowly that by the time it's done one spin, 243 Earth days have gone by. Meanwhile, it's positively whizzing around the Sun. Venus makes one full circuit of the Sun – completes a year, in other words – in 225 Earth days.

So, a day on Venus is a little longer than a year on Venus!

In a group of 23 people, there's a 50% chance two share the same birthday.

If you were to put 23 people in a room, there's a 50:50 chance of two of them sharing the same birthday. It may sound implausible, but it's fairly straightforward to demonstrate this with mathematics!

We need to start by calculating the total number of combinations there are between all 23 individuals.

Working one by one, and out of earshot of everyone else, the first person asks the other 22 people what their birthdays are. The second person then does the same, but can exclude the first person as they've already had that chat; this means they only have 21 comparisons to make. The third person asks all the group but can exclude the first two people, as they've already chatted, so they only have 20 people to check with. The fourth person then only has 19 to check with, and so on.

Adding up all possible comparisons $(22 + 21 + 20 + 19 + 18 + \ldots + 1)$, you get 253 combinations.

So, in total, there are 253 chances of matching birthdays between those 23 people. If we look at each person's birthday as one of 365 possibilities, that leaves 364 possibilities of it not being their birthday.

When comparing one person's birthday to another, in 364 out of 365 scenarios they won't match. This means that any two people have a 364/365, or 99.7%, chance of not matching birthdays.

We already know there are 253 comparisons, each with the same 99.7% percent chance of not being a match. When looking at the probability of two comparisons not matching, you multiply 99.7% by 99.7%. Do this 253 times over and you'll find there's a 49.9% percent chance that all 253 comparisons contain no matches.

To find out the percentage of matches, deduct 49.9% from one, which leaves you with 50.1%. Case closed.

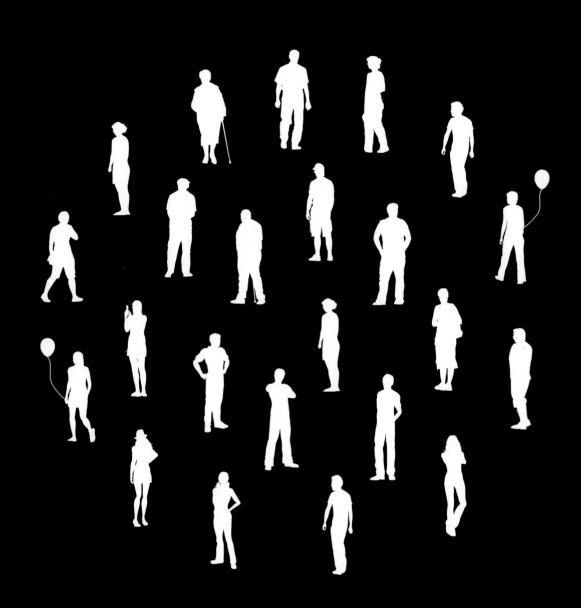

The tardigrade is the only animal that can survive in outer space.

Tardigrades, otherwise known as water bears, hold the unofficial title of 'world's most indestructible animal'. Don't search for them in a zoo, though, because they're micro-animals. They've been found everywhere, from the highest mountain tops to the deepest oceans, from tropical rainforests to volcanoes and the Antarctic – they really get around!

What these microscopic dudes lack in size, they more than make up for in all-round toughness and resilience. Boil them, deep-freeze them, crush them, dry them out – these 0.5 millimetre bears are almost impossible to destroy! As well as coping with extreme temperatures and air pressures, they can hang in there through air deprivation, radiation, dehydration and starvation. In conditions that would be almost immediately fatal to nearly all other life forms, tardigrades will survive and come back looking for more.

They can even survive in the cold, irradiated vacuum of outer space. A group of living tardigrades were sent into orbit on the outside of a rocket for ten days. When these minute astronauts returned to Earth, the scientists discovered that 68 percent of them had lived through their cosmic ordeal!

The secret to their amazing survival skills seems to be their ability to lose almost their entire water content, slow their metabolism down to near zero and enter a shrivelled-up state called desiccation. They can remain like this for decades.

Amazingly, once they find water again, within hours they're back from their dehydrated state, ready to face whatever challenges the universe can throw at them next.

Post-9/11 America has spent US$6,000,000,000,000 on wars, killing an estimated 500,000 people.

During the September 11 attacks of 2001, almost 3000 people lost their lives. Over two-thirds of the victims were civilians in this devastating foreign violation on US soil.

In the years since the strikes, the United States has led wars in Iraq, Afghanistan, Pakistan and Syria, costing taxpayers around US$5.9 trillion. These wars have also resulted in the deaths of an estimated 500,000 people. Of those, 370,000 died as a direct consequence of fighting related to US-led wars, and many more from destroyed infrastructure and subsequent lack of food and water. An estimated 250,000 of the victims were civilians, 14,500 were US military personnel and contractors, and 110,000 of fatalities are accounted for by enemy fighters.

Even more extraordinary is that an additional 500,000 deaths have been brought about by the civil war in Syria. They took place after the US government and other global powers attempted to remove the Syrian president in 2011.

The astronomical budget for these wars comes mainly from borrowing. Both the US budget deficit and the national debt have ballooned. Unless the debts are paid back by halfway through this century, the interest payments alone could total almost US$8 trillion.

Meanwhile, the wars rage on with no resolution in sight, and vast amounts of money continue to be spent, creating huge problems for both this generation ... and the next.

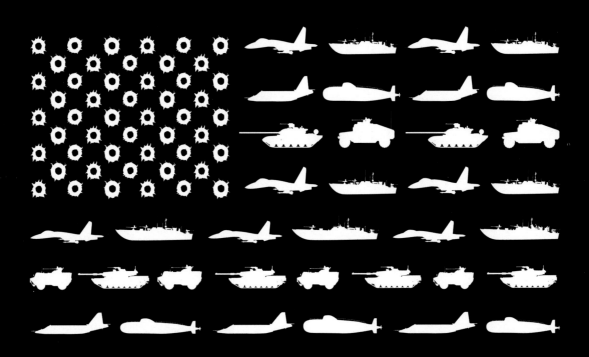

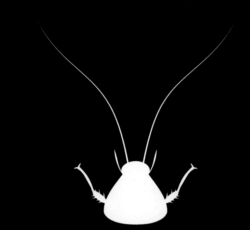
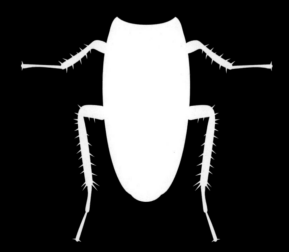

A headless cockroach is capable of living for weeks.

Cockroaches are downright gross. Try as we might to 'live and let live', there's just something about them that seems to give us the heebie-jeebies. While they're not attractive in either behaviour or looks, it's their zombie-like survival abilities that unnerve us more than anything.

Unlike us puny humans, cockroaches could make it through the radiation of a global thermonuclear war or being fully submerged in liquid for well over half an hour. What's grisly is that cockroaches can survive decapitation — well, for a week or so at least!

They achieve this mind-blowing feat thanks to their system of blood transfer. Cockroaches have an open circulatory system, which minimises pressure and, consequently, uncontrolled bleeding; clotting then allows their necks to seal off. Like all insects, cockroaches breathe through holes in their bodies, something for which no brain control is needed. Also, it isn't their blood that carries oxygen through their bodies.

Rather, they have set of tubes called tracheae that pipe air directly to their tissues. Being cold blooded, these creepers can get by with little or no food. Their last meal before decapitation will see them through for around a week. Talk about making their food last!

Even without its brain, the cockroach body can manage simple functions like standing, touching and basic movement. The cockroach body only dies because, without a mouth, it can't drink water and ultimately succumbs to dehydration. These headless cockroaches aren't operating anywhere near the capacity they are when their bodies are intact but, nevertheless, they are surviving.

What's even more terrifying is that the severed head can still survive too, waving its tiny antennae around until it eventually runs out of steam. Incredible and horrifying in equal parts.

Sufferers of hyperthymesia remember almost every life experience.

You might be able to remember that solitary goal scorer in the 1990 World Cup final, or maybe the name of the cutie you developed an instant crush on that day at the beach on a family holiday from your childhood. Most of us consider ourselves to have pretty good memories when we're able to recall the odd random event from our past.

But that doesn't seem impressive compared to the feats of those rare individuals who can instantly describe most of the memories from their entire lives.

These people have a condition called hyperthymesia, which means they're in possession of a superior autobiographical memory. Give them any give date in their personal history, and they can recount details such as what the weather was like and any significant events that occurred at that moment in time.

They can even tell you what they had for dinner that day! Show them a photograph from their past and they can immediately recall when and where it was taken, what they were up to and even what they were wearing.

They remember almost everything. And they forget almost nothing. How they're able to do this remains a mystery, but those with the condition describe seeing their memories as a series of distant pictures.

The phenomenon of mind-blowing recollection is a fairly new discovery. Until recently, it was believed humans couldn't possess such a remarkable kind of memory.

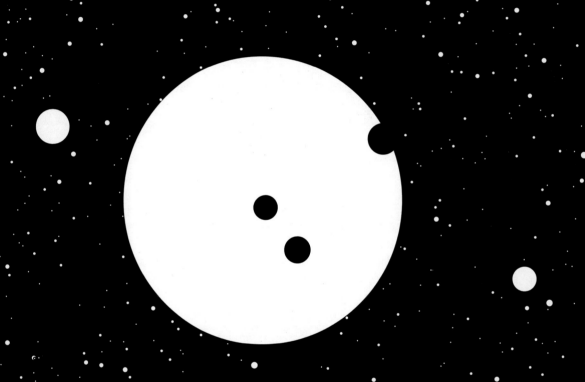

The existence of planets beyond our own solar system was only confirmed in 1992.

Here on Earth, we were late to the party when it came to verifying the existence of exoplanets, the name for planets outside our solar system. Now, however, we're making up for lost time as we contemplate thousands of new worlds – some of them possibly habitable!

This has really only been happening for a little over two decades. Astronomers located our first exoplanet in 1992 as it orbited a pulsar star. Next, in 1995, we confirmed the existence of an exoplanet orbiting a star similar to our Sun; the star was given the name 51 Pegasi.

Since then, we've discovered over 4000 new worlds in an amazing array of orbits and variety of sizes, in just a small portion of the Milky Way! If on average each star has at least one planet orbiting it, there could be a trillion planets in our galaxy alone!

The fact we can see exoplanets at all is quite incredible. Any planets out there only have a faint amount of reflected light visible from Earth because of the glare of the star it's orbiting. As a result, any planets seem very small and very dark.

For a clearer picture, we look at the stars themselves. If a slight wobble is detected, it indicates that a planet is orbiting the star. Orbiting planets cause gravitational pull, and these slight movements affect the star's normal light spectrum, or colour signature, altering what we can see from Earth.

Recently, we've also developed another approach. We have the technology to track stars at intervals and for regular amounts of time, noting any dimming – this reveals that a planet is orbiting. The dimming is then measured and recorded. The ultimate goal, however, would be to find a planet showing unmistakable signs of current life.

When the world's oldest person was born, a completely different set of humans occupied the planet.

If you're the oldest person alive, we can confidently state two things. The first is that you're a record breaker, so congratulations on that. The second is that all the humans on Earth have completely cycled over during your lifetime. Everyone who was on Earth before you were born is now dead; everyone currently alive on Earth was born after you.

On the other side of the coin, if you're the youngest person alive, you'd only hold this title for a little over four seconds as there's a current average of 250 children born every minute. That's more than 130 million in a year!

Looking back to the year 1800, there were a billion of us on the planet, with two billion by 1927. By 1987, this had more than doubled to a mind-blowing five billion – a number that has grown rapidly ever since.

This growth is all the more remarkable considering it took us nearly 2000 years to hit the billion mark, but only 200 years to get to seven billion. We're speeding up.

It's been estimated that since the dawn of humankind, a grand total of approximately 106 billion people have been born. To put this another way, the current living population accounts for roughly 6% of all the humans who've ever set foot on Earth.

Asia contributes almost 60% of the world's population. The overwhelming majority of this group live in China and India, which have roughly 1.4 billion inhabitants each.

Our global population is expected to hit eight billion in 2023 and ten billion by 2055. Earth is filling up, and it's getting crowded fast!

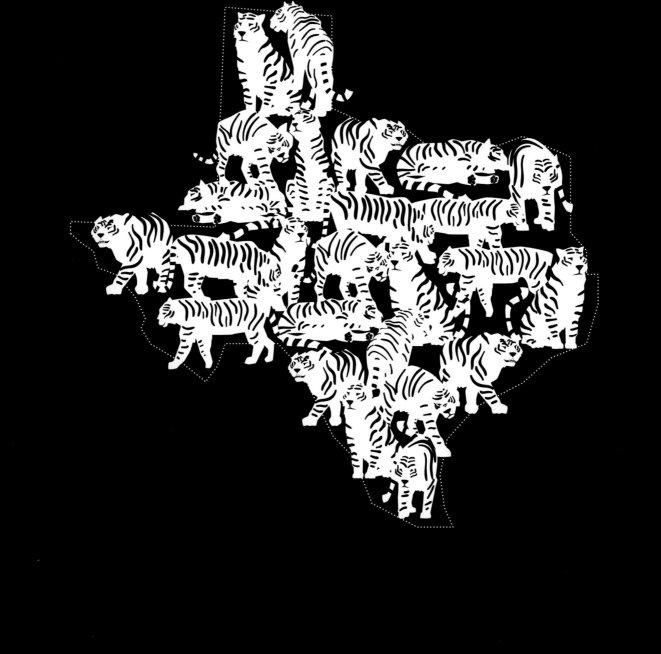

There are more tigers living in Texas than anywhere else in the world.

Awesome to look at and awesome in the way they behave, tigers are definitely the most stylish of the big cats. Sadly, these felines are endangered. Three of their nine subspecies have already become extinct and. of the remaining six, there are dwindling numbers. If we're not careful, tigers could disappear altogether.

Tigers are the largest of the cat species and can be found magnificently prowling around far east Russia, northeastern China, India, Malaysia and Sumatra.

And Texas.

As incredible as it sounds, in Texas it's easier to become a tiger owner than it is to become a dog owner – if the dog you've set your heart on has been labelled as dangerous. If you're a Texan who does happen to own a tiger, your state's hot climate is perfect for your pet: it can happily live outdoors throughout the entire year.

Because ranching land is relatively cheap, your extremely giant feline will be free to roam around your sprawling estate at their leisure.

It's estimated that there's somewhere between 2000 and 5000 tigers currently residing in the Lone Star State. There are also claims that this is a conservative guess and that the true figure is closer to 10,000! What makes this all the more mind-blowing is that, in the rest of the world, it's estimated that only around 3800 tigers are still living in the wild.

Many of the Texan tigers are kept in regulated animal sanctuaries and zoos, but unfortunately a lot are also kept in private homes and put on display in travelling zoos and roadside menageries. Not only is this cruel, but it's also hazardous. That's because captive animals still possess their natural instincts, and the possibility of an attack will always be there.

Mars is the only planet we know of exclusively inhabited by robots.

Martians are real and living on Mars right now. We know this is true because we're the ones who put them there. Humans have launched dozens of missions to Mars to better understand our cosmic neighbour. First we dispatched probes that flew by the Red Planet, gathering information in short bursts. Next were the longer lasting missions that orbited the planet for years, sending back detailed information through the depths of space.

Most challenging of all were the missions that landed on the surface and either stayed exactly where they landed or, even more astoundingly, set robots moving across the Martian surface.

The first of these rovers was NASA's *Sojourner*, a microwave oven-sized robot that trundled over Mars in 1997, covering 100 metres and spending 83 days in contact with us back on Earth.

NASA's next two rovers were more advanced. The two golf cart-sized rovers, *Spirit* and *Opportunity*, were both sent to Mars in 2004. While *Spirit* died in a sand dune in March 2010, *Opportunity* steadily explored for 15 years, covering over 45 kilometres. Sadly, a long period of radio silence following a monster dust storm in 2019 indicated that its mission was over.

At the time of writing, the car-sized rover *Curiosity* was the last of these robotic geologists to be deposited on Mars. *Curiosity* was programmed to collect data on Martian climate and geology, investigate environmental conditions favourable for microbial life and, most excitingly of all, conduct planetary habitability studies in preparation for possible human exploration.

The rich information *Curiosity* has sent back over its 20-kilometre adventure has seen the original two-year mission extended indefinitely!

SOJOURNER (ROVER)

Landed on Mars
July 1997

1997

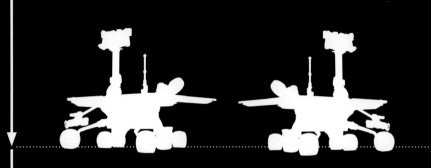

MARS EXPLORATION
ROVERS (MER)

Spirit & *Opportunity*
Landed on Mars
January 2004

2004

CURIOSITY (ROVER)

Landed on Mars
August 2012

2012

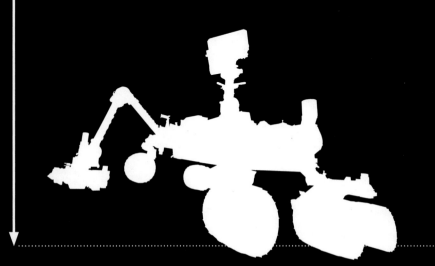

There was once a day without a yesterday.

It's hard for us to wrap our tiny minds around the idea that there was an origin point for all of space and time. Our brains hurt contemplating the universe's beginning roughly 13.8 billion years ago.

Now widely accepted, the Big Bang theory endeavours to explain how the universe developed from a microscopic hot point into the cosmos we inhabit today. The name of the theory is misleading as the universe actually expanded into being, rather than exploding. And we still have no clue what caused the creation of the universe, let alone what came before it or even what lies beyond it.

At the time of the so-called Big Bang, temperatures were way hotter than anything in our universe today. It was so hot that matter, as we observe and know it, could not exist. The universe expanded in a fragment of a second, becoming less dense and rapidly cooling down in the process.

Miraculously, it now measured 100 billion kilometres across and, even though it had cooled dramatically from its starting point, it was still a mind-blowing 10 billion degrees Celcius. Hands off!

As it developed, the elementary particles bound together to form protons and neutrons, the building blocks of atoms. Over the next few minutes, the expansion slowed down and the cooling continued. Around 400,000 years after the Big Bang, the protons and neutrons had joined to form the first atoms. By the time 600 million long years had passed since the Big Bang, the very first stars – similar to our own Sun – were born.

As you read this, the universe continues to expand. Not only is it getting larger but – mysteriously – it's also speeding up. It looks like our tomorrows are going to be every bit as mysterious as was our original day without a yesterday.

Ants have conquered the Earth.

As a species, humans do smug extremely well. We consider ourselves the dominant species on Earth. There are over 7.5 billion of us, and we've managed to cover the globe most effectively, building civilisations and engineering the land to suit our own needs.

But it turns out another species may already be in charge. Except they keep their dominance under our feet. Ants already control the planet. There are 10,000,000,000,000,000 of them – that's 10,000 trillion ants! This mind-blowing population works out at 1.3 million ants per human.

But our superior size surely means we're dominant, right? Wrong. If you compared the collective weight of every ant on the planet and the collective weight of every human, then these two biomasses are about equal. Not only are ants monumental in numbers, but they're everywhere too.

Given their ability to survive at the extremes of the natural world, it's no wonder almost every landmass on the planet has ants.

Then there's the fact that they're intelligent, *extremely* intelligent. Individually, a single ant may not appear to be at all that clever, but collaboratively ants can solve extraordinary problems with their hive mind. Like humans, ant colonies have the power to engineer the world around them. Building hugely elaborate interconnected chambers under the ground, these ant colonies work together, sharing jobs between them.

Humans have been on Earth for what seems like an eternity. But our roughly 200,000 years of existence have nothing on the ants, who shared the planet with the dinosaurs 130 million years ago – and survived the mass extinction that wiped those dinosaurs out! All hail our ant overlords.

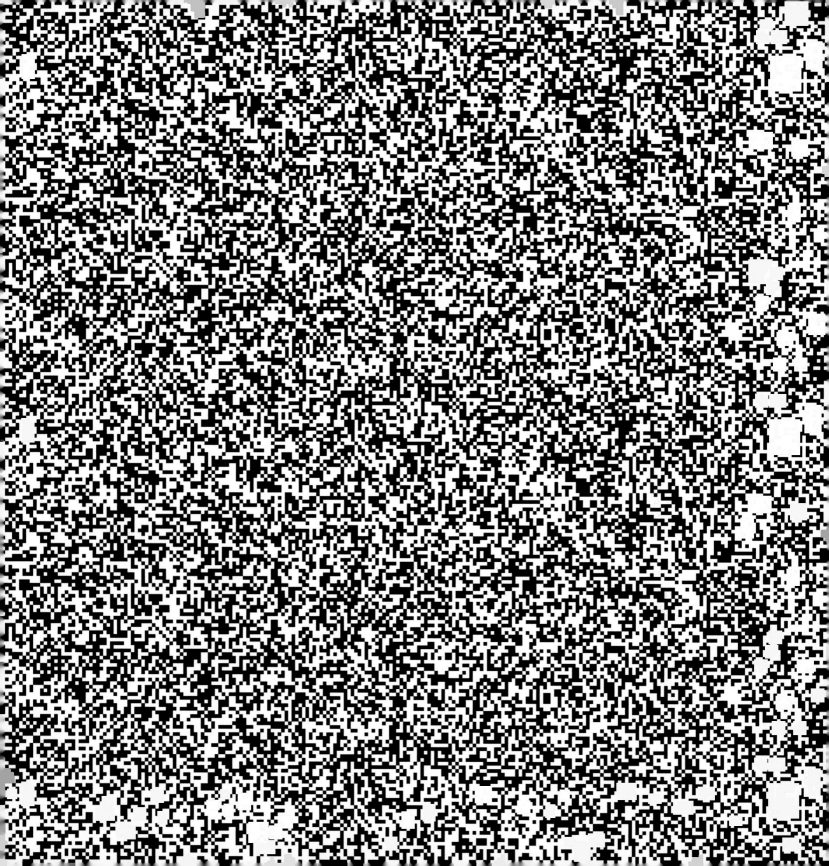

1% of the static on your TV is radiation left over from the Big Bang.

Here's something to try if you happen to revisit the past and find yourself struggling to tune your old analog television to a new channel. Stop mumbling profanities under your breath for a minute to contemplate exactly what that fuzzy stuff is distorting the picture on your screen. Yes, it's static, but did you know that 1% of it is caused by the afterglow of creation?

Commonly known as cosmic microwave background radiation, this static is residual heat from the Big Bang, the massive expansion from which the universe emerged 13.8 billion years ago.

It's the oldest light we can see, and it set out on its journey through space long before the Earth or our galaxy existed. This light is a relic of the universe's infancy, a time when it wasn't the cold dark place it is now but was instead a firestorm of radiation and elementary particles.

The familiar celestial objects that surround us today – stars, planets and galaxies – were formed when these particles coalesced as the universe expanded and cooled.

Cosmic microwave background radiation is critical to the study of cosmology as it bears the fossil imprint of those particles, a pattern of minuscule intensity variations from which we can decipher the vital statistics of the universe. It's like identifying a suspect from their fingerprint.

Next time you complain there's nothing on television to watch, remember that the white noise between channels means you can literally binge-watch the beginning of time and the origin of the universe.

Aliens have probably already observed us and concluded there's no sign of intelligent life on Earth.

In our boundless and ancient universe, with its billions upon billions of galaxies, it seems implausible, not to mention arrogant, to conclude that we simple apes are the only form of life enjoying its cosmic hospitality. As if! Perplexingly, however, we've found no proof otherwise.

There's a term for this contradiction between the lack of evidence and high probability of existence of extraterrestrial civilisations. It's the Fermi paradox, named after physicist Enrico Fermi.

We seem to need to be jolted out of seeing ourselves as the only thinkers in the universe. What if we're not as intelligent as we assume we are?

Generally, humans have an attitude of superiority towards the living creatures we share our planet with. We regard the insects, animals, trees and plants around us to be of a lesser intelligence. If we don't exploit them, more often than not we ignore them. They exist in the background of our lives.

Bear this thought in mind when you consider that any extraterrestrial civilisation with the technology to traverse deep space is likely to be considerably more intelligent than we are. And if they've already observed us — perhaps slipping in an undetected visit or watching us from afar — they probably arrived at much the same conclusion about us as we did about our fellow earthlings. Nothing to see here. No sign of intelligent life.

Sharks are older than trees.

Trees have long been associated with knowledge. It's said that the Buddha himself meditated under a tree, seeking its wisdom — and if the Buddha is coming to you for advice, you're doing something right.

Ancient plants started to make it out of the water 500 million years ago, and the first modern tree established itself in the Earth's developing forests around 370 million years ago.

We estimate there are over three trillion trees on the planet today. But, as is often the case, humans have been bad news for trees. Since our modern history began 6000 years ago, the number of trees on Earth has fallen by nearly 50%. Startlingly, 15 billion trees disappear every year.

Sharks are another of our planet's ancient life forms. Incredibly, they're even older than trees, though it's entirely understandable why the Buddha didn't go to one of them for advice.

Sharks may have been terrifying us for a long time, but they've been swimming in Earth's oceans for a lot longer still — 450 million years in total. That's 230 million years before the dinosaurs!

Fortunately for us, the thousands of teeth that shark produce during their lifetime are the perfect fossils, providing us with a record of their early evolution and history.

These prehistoric fish have lived through all five of Earth's mass extinctions. Ice ages, climate change, asteroids, toxic algae, you name it — sharks are born winners who've survived it all!

Nowadays, the shark's biggest challenge is us. Humans are slaughtering these ancient creatures at a faster rate than they can reproduce. If we're not careful, the shark's 450 millionth year might end up being its last.

If you removed the empty space in our atoms, the entire human race would fit into a sugar cube.

Everything you can see or touch – from your body, the planet we live on and every single star up in the sky – is made up of millions of billions of microscopical particles called atoms. They're the stuff of the physical universe, and if you paid attention in school you'd know that, in turn, atoms consist of protons, neutrons and electrons – and a lot of empty space. In fact, mainly empty space … It turns out that atoms are 99.999999999999% empty space!

You can visualise this empty space by taking an apple and picturing it as the nucleus of an atom. The relative size of the atom would be that of a cathedral.

If it happened to be a hydrogen atom, then it would have a single, fly-sized electron buzzing around its cavernous space.

So, if we're each mostly made up of empty space, imagine what would happen if we could remove it – what would actually be left?

Nada. Well, *almost* … The leftover mass of our body would be infinitesimally small, so small, that if we did the same thing to the entire human population, it'd be the size of a sugar cube.

But don't try putting this in your tea – it would weigh a few trillion kilograms.

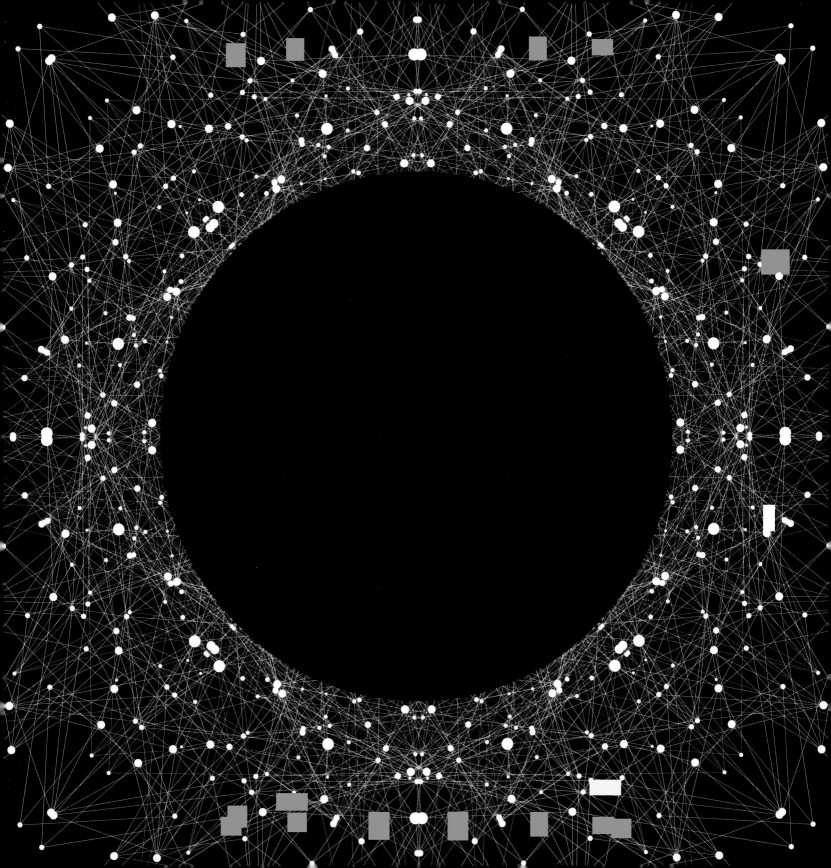

Monarch butterflies migrate 5000 kilometres annually.

Brown and orange wings with black and white spots are what mark the monarch butterfly out as the world's most recognisable lepidopterous insect. They're often seen fluttering gracefully in our back gardens, looking as if they might fall to the ground with exhaustion at any moment.

But looks can be deceiving. These delicate creatures have real staying power. With a wingspan of just 10 centimetres, and capable of flying at speeds of up to 40 km/hour, these butterflies undertake one of the most spectacular migrations on the planet. It's an epic journey from which they will never return.

Every autumn, millions of North American monarch butterflies travel up to 5000 kilometres from Canada and the north of the United States, down through the country and all the way across the border to southern Mexico. They undertake this long and arduous journey to escape the cold winter of the United States.

Spookily, each year's cohort unhesitatingly navigates the identical path, even though they've never made the journey before. Their internal compasses point them in the right direction and off they go!

The monarch butterflies use some of the same energy-saving strategies as migrating birds. They take advantage of any updrafts of warm air, or thermals, to glide through for a while and give their tiny wings a rest.

Once they arrive in the cooler climate of southern Mexico, they set up camp in exactly the same trees used by other butterflies in previous migrations. The females will lay their caterpillar eggs, and the cycle goes on until a new generation of butterflies starts the epic return migration north.

There's a hexagonal cloud pattern around the north pole of Saturn that's bigger than Earth.

Just when you thought Saturn and its stunning rings couldn't get any more amazing, along comes the mind-blowing discovery of a giant hexagonal cloud hovering at the top of the planet. Saturn, you've done it again!

The incredible geometric cloud pattern was first discovered by NASA's *Voyager* during its fly-bys in 1981, and confirmed thanks to recent close-up observations made by NASA's *Cassini* spacecraft. This six-sided cloud has a churning storm at its centre and is about 32,000 kilometres wide and 100 kilometres deep!

From Earth, astronomers then set about capturing images of the hexagon and studying its constantly moving pattern of swirling clouds. They noticed that, over a couple of years, it gradually changed colour from blue to gold. It's possible that seasonal variations in sunlight caused the change.

We have no definitive explanation for why the cloud pattern is there. One theory is that the cloud may be a jet stream made of atmospheric gases moving at 320 km/hour. The hexagonal shape is the real puzzle. It's rare to see straight lines in natural settings. Plus, our storms here on Earth are circular and never contain any angles.

Deepening the mystery, it has now been found that a second massive hexagonal vortex is floating high above the storm. This vortex sits about 290 kilometres above Saturn's upper atmosphere.

Frustratingly for scientists studying these cosmic anomalies, Saturn's seasons last for roughly 30 Earth years, making for long winters. This means we have a while to wait before we can study these summer seasonal vortexes again.

12,700 km

32,000 km

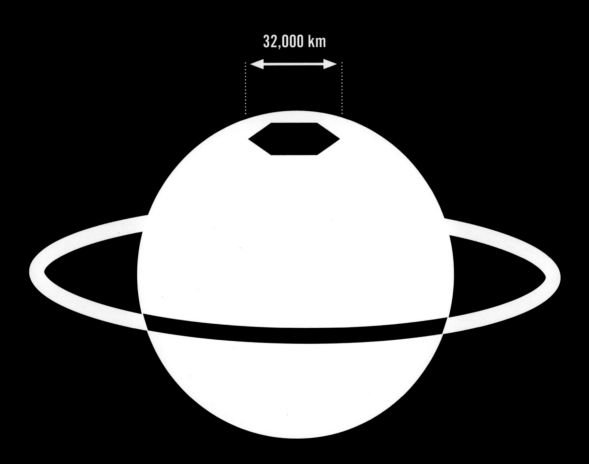

Free will is merely an illusion created by our brains.

All of us go about our lives convinced we're making our own conscious choices. We make around 35,000 decisions every single day: taking that particular pair of jeans out of the wardrobe and leaving the white ones, or having a large coffee this morning as opposed to a regular-sized cup. Not counting the time we're asleep, we make roughly 2000 decisions per hour, or one every two seconds.

Or so we think.

Because our behavior is made up of such a complex and disorderly mix of factors, some neurobiologists reject the idea that there's a singular 'you' making your decisions. They argue that the decisions we make are shaped by a staggering number of factors.

Things like our mother's experiences while we're in the womb, our genes and hormones, our parents' behaviour when we were growing up and whether we witnessed any violence in childhood.

To a neurobiologist, every human action is caused by preceding events, and that includes events in the brain. Our experiences are inescapable. You think you can freely choose to do one thing or another? No chance. There can be no such thing as free will.

The concept of free will probably arose to give us a feeling of control over our lives. Our belief in free will also gives us licence to punish those who do wrong to others. But if the neurobiologists are correct, the criminal justice system will need a radical review, as those we punish are simply pre-programmed biological computers without any agency.

Everything currently living on Earth originated from non-living matter.

Calculating exactly how many species exist on Earth is a tough task. Things have gotten just a wee bit out of hand, and it makes taking a census slightly tricky ...

We're not entirely sure how many animals there are wandering around, let alone the numbers of plants, fungi or microbes. Life is both incredibly abundant and diverse. This is mind blowing when you consider that life arose from nothing more than the water, atmosphere and rocks of Earth's early environment. And all of this in only a few hundred million years. The mere blink of an eye in cosmic chronology.

The origin of life on Earth is a scientific mystery that we are yet to, and may never, solve. One way to account for the origins of life is called abiogenesis, the evolution of inorganic molecules to organic molecules.

This is the theory that the metamorphosis from non-living matter to living entities wasn't a solitary event, but rather a slow-moving process of increasing complexity.

There are many different concepts within abiogenesis, most of which are linked to the idea of a 'primordial soup'. This supposes that the early Earth had a chemically reducing atmosphere which, when exposed to energy in various forms, produced simple organic compounds that accumulated in a kind of soup. This may have occurred at ocean shorelines or underwater hydrothermal vents.

Crucially, there was further transformation, more complex organic polymers mutated and, ultimately, early life began to develop in the chemical soup.

In other words, life had begun. And it wouldn't be long before it began looking for answers as to why.

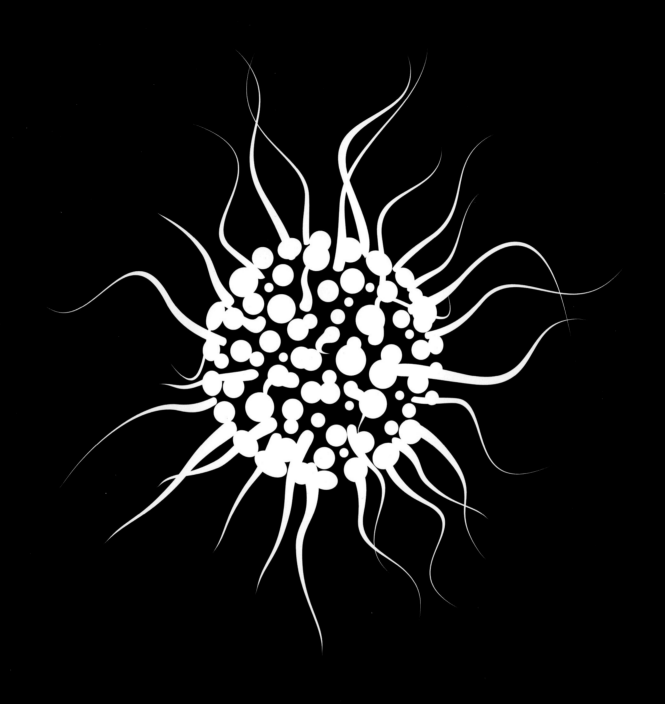

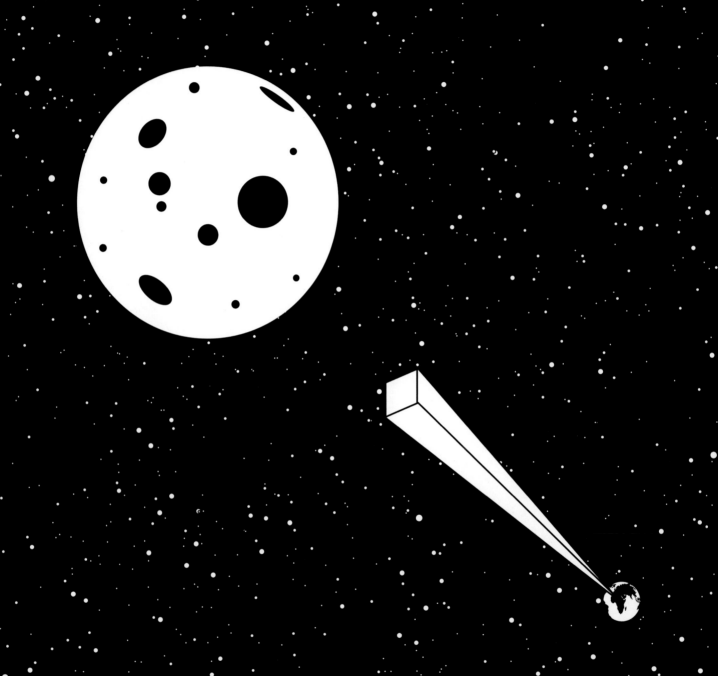

If you could fold a piece of paper 42 times, it would reach to the Moon.

Anyone familiar with making paper planes knows that an ordinary sheet of paper folded in two instantly become twice as thick as it was before. Devastatingly simple. Have you ever wondered how many times you'd need to fold this paper to reach the Moon? Probably not, but you're about to find out!

The Moon is 384,000 kilometres away. It's easy to assume that's a lot of folds – thousands, possibly millions. The answer, rather astonishingly, is just 42 times!

And it's all because of the mind-blowing power of exponential numbers. One sheet folded becomes two; you fold that again to make four; give it another fold and it becomes eight; do it again and it's 16. By the sixth fold, it's 64 sheets thick. In reality, there's no chance you could fold your way to the Moon with just a standard sheet of A4.

This is an imaginary piece of paper, one you can continue folding as much as you like. If you continued folding your magic paper, the seventh fold would produce 128 sheets; the eighth fold 256 sheets; and by the ninth fold, the paper would be 512 sheets thick. This is the depth of a standard ream of paper.

By the time you've folded your paper 20 times, the pile would be over 10 kilometres high – taller than Mount Everest! Twenty-four folds and your pile will hit outer space. The Moon is now clearly in sight, and after just 42 folds, you'd have reached your destination. In fact, you'd have passed it!

This is all thanks, once again, to exponential numbers – turning small things into massive things by doubling what you have again and again and again and again.

Snowflakes always have six sides.

When you're warm and snug, snow falling in winter is quite the sight. Fluffy white flakes serenely floating down from the sky, covering all that's below in a lovely white blanket.

The magic intensifies when these flakes are viewed through a microscope. Here, mind-blowing crystalline patterns are revealed in all their geometrical glory. If scientists could convince poets – which is extremely unlikely – the term 'snow crystal' would replace 'snowflake'.

Behind this beautiful soft snow is some beautiful hard science, explaining not only why two snow crystals are never the same, but also why they always have six sides!

A snow crystal forms when water vapour freezes around a small particle, like dust or pollen.

The molecular structure of water – one oxygen and two hydrogen atoms – dictates that the pattern these new snow crystals begin to form is a basic hexagon, and all snow crystals have a six-sided shape. Ta-dah!

But that's where the similarities end, as every snow crystal that's ever fallen to Earth is completely unique. Variations in humidity and temperature are what causes every snow crystal to form its individual shape. Low humidity generates lots of simple plates and simple hexagonal blocks, whereas with higher humidity you get more branched structures.

Even if you found two snow crystals that appeared identical, at the atomic level they'd look entirely distinct. Every snow crystal contains 1018 water molecules, meaning there are more possible variations for their arrangements than there are atoms in the universe! You won't be seeing twin snow crystals anytime soon.

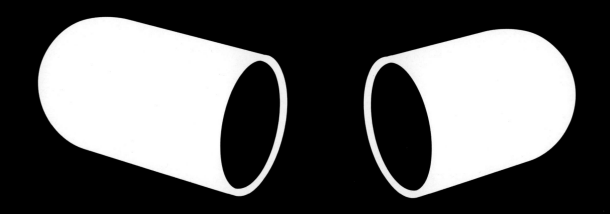

The placebo effect is all in your mind. And it's real.

The placebo effect is the merging of biology and psychology that gives us superhero-like healing powers and is, arguably, one of the most interesting phenomena in science.

The classic example of the placebo effect is when a patient believes they're receiving medication but is really taking a sugar tablet. Then they experience a miraculous improvement in their condition. Scientists cannot fathom how it works, only what it reveals: the placebo effect shows that our minds have far more control over what happens in our own bodies than we may like to think.

It's believed that the improvement derives from the patient's expectation that the placebo will help them. It seems that the greater the belief a person has that the treatment is going to alleviate their condition, the greater the chance they'll experience a positive change.

Placebos often come as an inactive substance like a sugar pill, saline solution or distilled water. Pharmaceutical researchers use them in double-blind trials to help them better understand what effect a new drug or treatment might have on a medical condition. Broadly speaking, they administer placebos to some groups and new drugs to others. They measure and compare the effects on participants.

The placebo effect has been extensively studied. It's been shown to work on all sorts of people, regardless of their gender or background, and to affect conditions such as depression, pain, sleep disorders, irritable bowel syndrome and menopause.

Actual physical changes can occur thanks to the placebo effect. Researchers have observed increases in the body's production of endorphins, one of our natural pain relievers.

It rains diamonds on Jupiter and Saturn.

Along with Neptune and Uranus, Jupiter and Saturn are the gas giants of our solar system. (If you need to pause to guffaw at the combination of 'Uranus' and 'gas' in the same sentence, please go right ahead.)

Jupiter is larger than Saturn, but both are way bigger than Earth, made primarily of hydrogen and helium, each have over fifty moons. Most impressive of all is the fact that on Jupiter and Saturn, it rains diamonds. Crazy, but true!

It may sound like a Prince album track but the diamond rain on Jupiter and Saturn has some marvellous cosmic chemistry behind it.

On Earth, deep below the surface of our planet, carbon atoms bond together and, over billions of years, start forming crystals. Things are done a little differently on Jupiter and Saturn.

Both planets have huge amounts of methane gas in their atmospheres, which turns to soot when struck by lightning. Soot is impure carbon particles, and when this carbon falls from the atmosphere, the intense heat and gas pressure turn it from chunks of graphite into diamonds. Sky bling.

Millions of kilograms of diamonds are created this way every year on Jupiter and Saturn, so it's doubtful they'd be regarded as precious stones there. But, with the largest diamonds coming in at a centimetre in diameter, you could definitely put a ring on it!

As they fall through the extreme heat and pressure of the atmosphere, they can no longer remain solid, resulting in liquid diamond rain. Another Prince B-side.

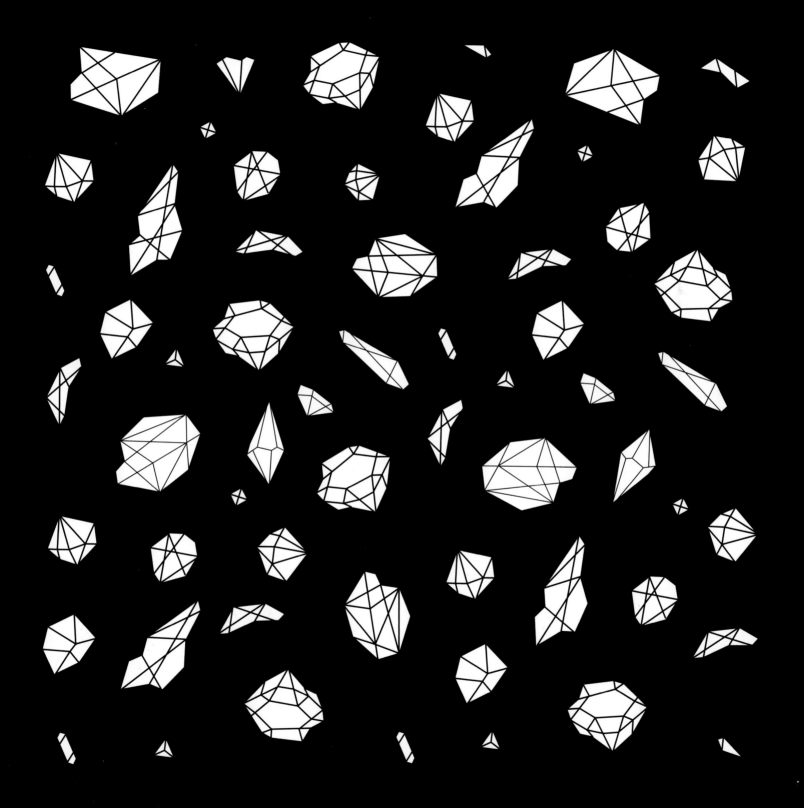

PERMIAN PERIOD
225 million years ago

PANGAEA

TRIASSIC PERIOD
200 million years ago

LAURASIA

GONDWANALAND

JURASSIC PERIOD
150 million years ago

CRETACEOUS PERIOD
65 million years ago

PRESENT DAY

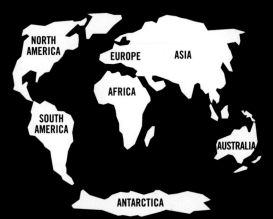

NORTH AMERICA

EUROPE   ASIA

AFRICA

SOUTH AMERICA

AUSTRALIA

ANTARCTICA

Earth used to consist of one supercontinent surrounded by a superocean.

The Earth we see today didn't always look this way. Our planet has had more than one stunning makeover in its 4.5-billion-year history. As a result of tectonic plates shifting and churning beneath the Earth's crust, several supercontinents have formed and broken.

The most recent supercontinent existed 335 million years ago. At that stage, all the land on the planet was one single giant mass surrounded by a single body of water. It would have looked incredible! This landmass was called Pangaea, and the magnificent superocean surrounding it was called Panthalassa.

Pangaea first began to break apart about 200 million years ago, when Gondwana (Africa, South America, Antarctica, India and Australia) separated from Laurasia (Eurasia and North America).

When Gondwana split up — about 150 million years ago — India peeled off from Antarctica, and Africa, and South America drifted. About 65 million years ago, North America broke away from Eurasia, leaving the continents configured as they are today.

In Earth's history, supercontinents have formed many times, only to fracture into new continents. Currently, Australia is inching toward Asia, and the eastern side of Africa is slowly pulling away from the rest of the continent.

Earth really likes to mix it up, making it pretty certain that the current continental arrangement is unlikely to be the last. Within the next 250 million years, Earth will be going retro. The Americas and Africa will merge with Eurasia to form a supercontinent almost as large as Pangaea was originally!

If America's prison population were a city, it'd be one of the country's largest.

Nobody does incarceration quite like the United States does incarceration. According to publicly available figures, America locks up more of its citizens than any other country on Earth. More than bigger countries, like China and India, and more than those with supposedly tougher regimes, like Russia and the Philippines.

Currently, there are nearly 2.2 million individuals being held inside prisons across America. If all these people were moved into one correctional facility, you'd have one of the largest new cities in the whole country, bigger in population than Philadelphia or Dallas!

The large size of this incarcerated population is a fairly recent development. In 1980, around 500,000 people were jailed.

Over the next 40 years, give or take, the prison population more than quadrupled, doubling in size during the last two decades alone.

Look at different aspects of these statistics and you'll see that for every 100,000 people in the United States, 724 of them are in a jail cell. And get this — America accounts for roughly 5% of the world's population but has around 25% of the world's prisoners. More people are behind bars there than in the top 35 European countries combined.

Keeping all these people locked up isn't cheap. The running costs of the US prison system is roughly US$80 billion per year. If you take into account the social cost of incarceration to families and communities, that figure jumps to over a trillion dollars.

NEW YORK 8.6 million

LOS ANGELES 4 million

CHICAGO 2.7 million

HOUSTON 2.3 million

US PRISON & JAIL POPULATION 2.2 million

PHOENIX 1.6 million

PHILADELPHIA 1.6 million

SAN ANTONIO 1.5 million

SAN DIEGO 1.4 million

DALLAS 1.3 million

AMERICA IS HOME TO 5% OF THE WORLD'S POPULATION, BUT 25% OF THE WORLD'S PRISONERS.

US PRISON & JAIL POPULATION

REST OF THE WORLD'S PRISON & JAIL POPULATION

Cicadas use prime numbers.

If you've ever lain in bed during a midsummer's night, unable to sleep thanks to the cacophonous din of cicadas, then you're probably not a fan of these diminutive flying insects.

Their life cycle is a strange one as they spend most of their days underground and when they do finally emerge, there's no time for pleasantries as they quickly find a mate, reproduce and die.

Annual cicadas make this appearance every year, but periodic cicadas stay underground for much longer, appearing at intervals of seven, 13 or even 17 years.

Certainly, the length of the period cicada's underground existence is curious. But why would each of these intervals be a prime number, meaning they're only divisible by themselves and by one?

Could that be a coincidence? For a long time, we thought so. But now it seems that these insect mathematicians have been using prime numbers to their advantage. And all this without a calculator!

For instance, cicadas with a 13-year life cycle and ones with a 17-year life cycle will hardly ever meet. When both species of cicada do come out together – once every 221 years (13 multiplied by 17) – the numbers will be massive. So it's handy that it's rare for them to have to compete for the same food.

Another benefit of the strategically irregular gaps is that they prevent predators syncing up with the reproductive cycle of the cicadas. A predator with a three-year life cycle would only meet up with the 17-year periodic cicadas once every 51 years (three multiplied by 17).

The pyramids of Giza were once covered in hand-polished, highly reflective white limestone. With a gold capstone.

A declaration that the pyramids of Giza are awe-inspiring won't exactly elicit a gasp. Because of course they are! This trio of enigmatic monuments from a long-gone civilisation tower above the desert landscape. A picky visitor may observe sorrowfully that the giant stone steps have been blackened with smog and pollution, but there's no denying that the pyramids remain a powerful sight.

The grandest of these is called the Great Pyramid. It took up to two decades for it to be completed, and for nearly 4000 years it stood as the tallest man-made structure in the world. Without doubt, the Great Pyramid is the best known relic of ancient Egypt. Pity we still aren't sure how it was actually constructed or what it was built for.

What's impressive, though, is that all three pyramids of Giza were originally encased in smooth, intricately cut, hand-polished, highly reflective white limestone. They would have shimmered spellbindingly and looked like colossal light sculptures dropped into the desert from the sky. But wait, there's more: for added bling, a giant gold capstone adorned each peak. Stunning!

Imagine, back in the day, the magical moment you first set eyes on the port city, bathed in sunlight and basking in the ethereal glow of these mind-blowing monuments.

Lamentably, between earthquakes and the need for mosque-building materials, all the casing stones were spirited away centuries ago, robbing today's tourists of the most spectacular selfie ever.

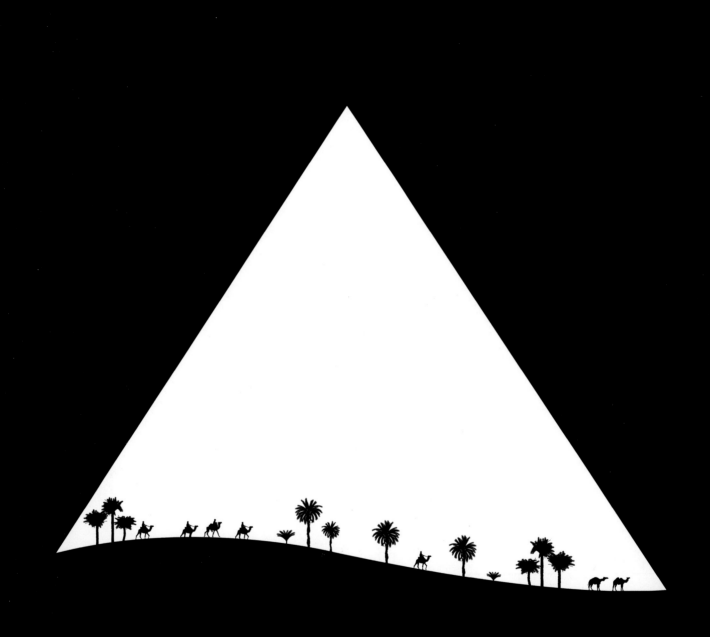

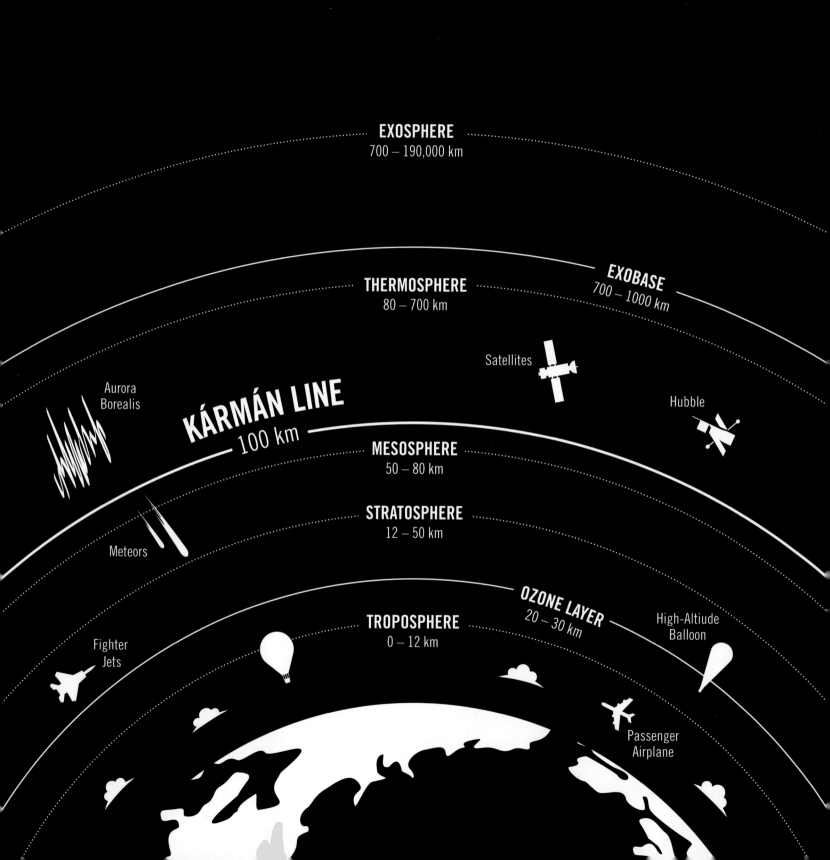

EXOSPHERE
700 – 190,000 km

THERMOSPHERE
80 – 700 km

EXOBASE
700 – 1000 km

Satellites

Hubble

Aurora
Borealis

KÁRMÁN LINE
100 km

MESOSPHERE
50 – 80 km

STRATOSPHERE
12 – 50 km

Meteors

OZONE LAYER
20 – 30 km

TROPOSPHERE
0 – 12 km

High-Altiude
Balloon

Fighter
Jets

Passenger
Airplane

Outer space is only an hour's drive away.

Picture yourself driving along the freeway, music playing as you cruise at a cool 100 km/hour. Now imagine the freeway going straight up. In seven minutes, you'll have driven past all commercial airliners. In eighteen, you'll have left behind the ozone layer. After half an hour, you'll have an astronomical view of any meteors, burning up as they hit the stratosphere. At the one-hour mark, you'll reach the Kármán line, the boundary between Earth's atmosphere and outer space, 100 kilometres above sea level. Now you're at the very edge of space, the point at which an aircraft can no longer achieve any lift.

To push past the Kármán line and continue the journey into space, your car would need to have enough thrust to leave Earth's orbit. You'd need to hit a mind-bending speed – around 29,000 km/hour – a task usually achieved with the help of a rocket.

For our purposes, let's pretend that a ridiculously expensive car customised for space travel could do the job too.

As you level your trajectory your car will now be orbiting Earth, and you'll be able to spot the Hubble Telescope floating above as it peers out into the cosmos. The Aurora Borealis will be a spectacular light show below you.

The most spectacular view of all will be Earth's atmosphere, an incredibly fragile pale blue line. At 100 kilometres up, you'll notice how our atmosphere grows thinner and thinner as it eventually fades into the vast emptiness of space. Time to take your hands off the wheel, put some Bowie on the stereo, sit back and enjoy the ride.

Trees can secretly communicate with one another.

There are over three trillion trees on the planet. That's at least 400 trees to every single one of us. When was the last time you looked at a forest and gave some proper thought to what it is you're seeing? Trees have existed for aeons – around 300 million years – much longer than us. It might be time to start looking at them a little differently …

Whereas once we viewed trees as direct competitors of one another – seemingly battling for access to water, light and space – there's a growing realisation that we've had this wrong. Trees are far more aware, intelligent and social than we previously thought. Trees in a forest have evolved to exist in reciprocal and harmonious relationships and they have a collective intelligence similar to that of insects. They support one another and have an interest in keeping each member of their community alive.

Trees are social beings, with forests being their way of networking socially. Away from human eyes, their underground root systems connect them all, allowing these arborial buddies to communicate with one another. A wood-wide web.

So, what exactly are they communicating underground and beneath the bark? When they're under attack – from humans who cut their roots or by insects that eat them – trees communicate their distress in electrical signals along their roots and across fungi networks to other trees nearby. Using the same root communication system, they're able to feed stricken trees, nurture young saplings and bar trees from outside their group in order to keep their own community strong.

Trees are just like us after all.

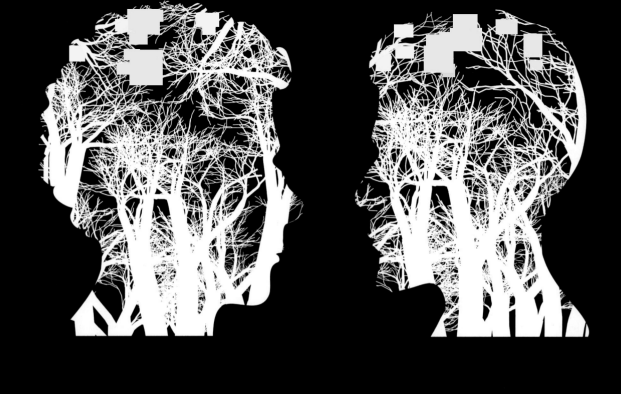

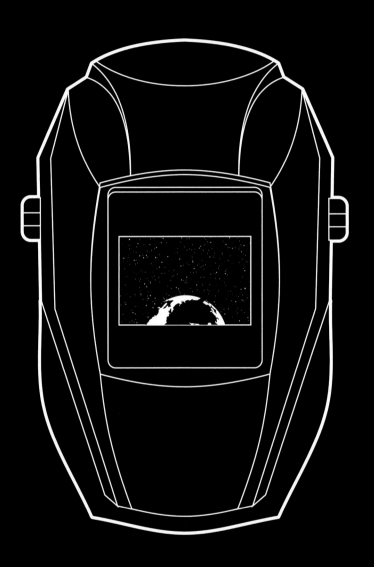

Two similar metals touching in space will permanently fuse together.

If an astronaut floating in space were to touch two metal rods together, they'd instantly bond into one single rod. This is known as 'cold welding'. The explanation for this mind-blowing phenomenon is that in the oxygen-deprived vacuum of space, atoms that come into contact with one another have no way of knowing that they're actually in different pieces of metal. To them, they're all one big similar-looking family. An extremely close one.

By contrast, wherever metal comes into contact with oxygen, it catalyses into a chemical compound called a metal oxide. So, on Earth, metals are almost always covered by a thin film of metal oxide, making full metal-to-metal contact extremely rare.

In space, there's no atmosphere to provide oxygen and therefore no way to renew this metal oxide coating. When the metal oxide wears off and the metals press together, hey presto, there's total metal-to-metal contact.

Here on Earth, welding metals together is anything but a cold process. It involves so much heat that sparks fly as the external coating burns away to let the raw metals fuse with each other into one element.

A process similar to cold welding can be tried out in your own kitchen. Though not spectacular, you can experience it without the expensive and possibly life-threatening trip to space. If you take a couple of ice cubes, let them warm up a bit, then press them together, they'll weld into a single lump.

90% of the world's population lives in the Northern Hemisphere.

We all know of the Equator, the imaginary line that separates the north of the Earth from the south. But did you know that nearly everybody on the planet lives to the north of it?

There are many reasons the north is the place to be – mainly geographical and practical ones. The area of land between the Equator and the North Pole is much greater than the area of land between the Equator and the South Pole. In fact, the Northern Hemisphere accounts for 68% of the Earth's total landmass. And where there's more landmass, there are generally more people. Take a bow, the densely settled nations of India and China. Together, these two are home to a whopping 36% of the world's human inhabitants.

Below the Equator lies the continent of Antarctica, which – while extraordinary – is a real estate agent's nightmare. Located in the south polar region, it is permanently covered by thick ice sheets. Nobody except penguins and seals thinks it's a good idea living here.

The Southern Hemisphere also contains Australia, which is mostly desert. Consequently, it boasts a small population compared to other developed countries.

By contrast, the land in the Northern Hemisphere is much more temperate, meaning more of it can be dedicated to large-scale agriculture.

The land in the Southern Hemisphere also has limited overland connections, whereas the Northern Hemisphere is heavily linked, which encourages long-range commerce. This is why many more major civilisations have risen up in the Northern Hemisphere than in the Southern Hemisphere.

As a species, we need the life-sustaining elements of land, water and air. All three of these are more readily available in the Northern Hemisphere than in the Southern Hemisphere.

EQUATOR

Octopuses have three hearts, nine brains and blue blood.

Even though no-one knows what an extraterrestrial being looks like, we often call octopuses aliens. With their large bulbous heads, amazing tentacles and impressive shape-shifting abilities, they do have an otherworldly look. There have been theories that these eight-limbed cephalopods either arrived here aboard comets as fertilised eggs or that they evolved from squids after the introduction of alien DNA.

Octopuses possess some wild physical differences to us. For starters, they have three hearts. Two move blood beyond their gills, while the third keeps circulation flowing for their organs. One brain seems a bit boring — how about nine instead? Perhaps that's why they're such extremely intelligent creatures.

Octopuses have a central brain in the head, which controls their nervous system, and one small brain in each of their eight arms. Their arms literally have a mind of their own and work independently of each other, yet they always operate with the same purpose. Two-thirds of an octopus's neurons reside in its arms, so these guys can take multi-tasking to the next level!

If you're going to be different, why stop at hearts and brains? Red blood is so yesterday; blue blood is where it's at! Not only does it look fabulous, it's more efficient at transporting oxygen around the body at low temperatures. Blue blood allows octopuses to live near the sea floor, often in extreme or arctic temperatures.

Our blood is red because it contains iron, but theirs contains copper, which when mixed with oxygen results in the blue colour of their blood.

Uranus rotates on its side.

Uranus is best known as the planet in our solar system most likely to illicit a little snigger. But there's much more to the seventh planet from the Sun than some good old-fashioned potty humour.

In our solar system, Uranus is the third largest planet by diameter and by mass. It's almost four times larger than Earth and, like Saturn, has its own set of mysterious rings – not to mention it also has 27 moons, which sounds downright greedy!

The most amazing thing about Uranus, though, is that for some bizarre reason it rotates on its side. The planet is on such a big tilt that it looks as though it took a lie-down one day and decided to stay that way.

All planets have some form of tilt – Earth is angled 23.5 degrees away from the Sun's equator; Mars is at 25 degrees and even Mercury is tilted 2.1 degrees. Planets love to tilt. But with its 97.8-degree lean, Uranus is just showing off!

It's probable that, in the distant past, Uranus suffered a massive cosmic collision that pushed it over. What's even more likely is that it endured multiple blows. A single, massive collision would have flipped the planet over entirely, leaving it rotating in the opposite direction. But it would have taken many collisions to stop the continuous roll and leave Uranus with its current configuration.

Planetary pinball!

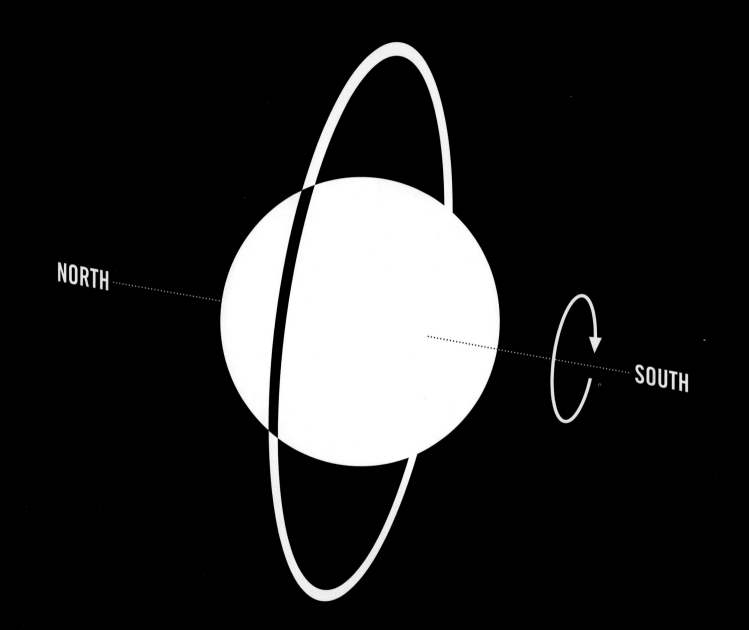

If Earth were the size of a basketball, the life-supporting atmosphere would be a thin sheet of plastic around it.

Extending from the surface of Earth to the edge of space is our atmosphere, a layer of life-preserving gases — commonly known as air — that surrounds the planet. It's held there by Earth's gravity, the same force that keeps us anchored to Earth. Our atmosphere contains 78% nitrogen, 21% oxygen, 0.93% argon, 0.04% carbon dioxide, and small amounts of other, less important, gases.

While we think of the atmosphere as a massive film of air around us, in reality it's exceptionally thin relative to the size of the Earth. In concrete terms, Earth is 12,742 kilometres in diameter and the thickness of its atmosphere is about 100 kilometres. If we shrank the Earth down to the size of a basketball, the atmosphere would only be as thick as a sheet of thin plastic.

Disturbingly, some of its atmosphere gets lost to space. But because of Earth's substantial size, our gravitational pull is strong enough to retain most of our precious air. A heartfelt thanks to our planet for that!

Just as well Earth is definitely not as small as a basketball because then the grip of its gravity would be dangerously weak. We'd even be in trouble if ours was a smallish planet like Mars. This size issue is one of the reasons Mars lost most of its original atmosphere and turned into the dusty red neighbour we know today.

The time people have spent playing Call of Duty is longer than the entire span of human history.

A major part of our evolution as a species has been the development of opposable thumbs. While they once allowed us to make primitive tools and weapons, we now use them to press little plastic buttons on video game controllers. And we do a lot of pressing …

The Homo genus — including the human species, *Homo sapiens* — came onto the scene more than two million years ago. *Homo sapiens* were distinguished from preceding species — none of which survived — by our bigger brains, better tool-making skills and ability to travel far beyond our African origins.

While our ancestors date back more than six million years, modern humans only evolved about 200,000 years ago.

From these humble origins, we've come a long way, to the point of developing video games that have become a multi-billion-dollar industry, played by over a billion people worldwide.

Call of Duty, the first-person shooter franchise, was released in 2003 and fast became a global phenomenon. Collectively, *Call of Duty* has had players glued to their consoles for approximately 2.85 million years of playtime.

That's 25 billion hours — longer than modern humans have existed on Earth! Players cumulatively spend the equivalent of 1900 years playing *Call of Duty* a day. Every single day.

And this is just one of over 50,000 video games in existence, with new games being released daily!

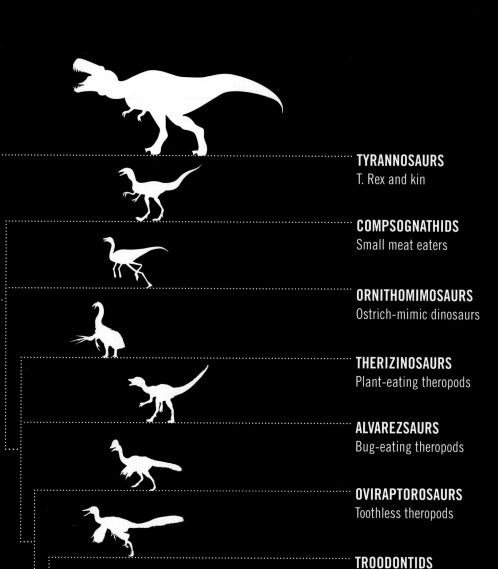

TYRANNOSAURS
T. Rex and kin

COELUROSAURS
Derived theropods

COMPSOGNATHIDS
Small meat eaters

ORNITHOMIMOSAURS
Ostrich-mimic dinosaurs

THERIZINOSAURS
Plant-eating theropods

ALVAREZSAURS
Bug-eating theropods

OVIRAPTOROSAURS
Toothless theropods

TROODONTIDS

Birds are dinosaurs.

Dinosaurs dominated the Earth for 165 million years – that's one hell of a long run! But as we know, all good things must come to an end, and for the dinosaurs that end came 65 million years ago. It's likely a giant asteroid rudely smashed into Earth, darkening the planet for months and transforming it into a burning apocalypse. Bye bye, dinosaurs; hello, extinction.

Well, perhaps not quite …

Life somehow found a way, and the small and carnivorous theropod dinosaurs survived the mass extinction. The original battlers!

On went the march of evolution … Fast forward to the discovery of the first *Archaeopteryx* fossil. What distinguished it from other theropod fossils of that era were the clear imprints of feathers around its body.

Not only was this fossil glamorous, but it provided the first evidence of shared features between modern birds and theropod dinosaurs. The evolutionary link between dinosaurs and birds had been established.

Further investigation of this bird–dinosaur link prompted a search for traces of other feathered dinosaurs. Sure enough, in the 1990s, some were uncovered in China – their bones were very light and scientists deduced that they'd laid eggs, all things in common with modern birds.

Modern birds not only evolved from dinosaurs, but they've thrived while doing so, evolving into 10,000 different species!

For the closest thing you'll get to a dinosaur encounter, take a good look at a chicken. Its DNA has changed less over time than other avian species. If you see one eat a live mouse, it'll all make sense … Chooks are freaks.

There are more stars in the universe than grains of sand in the world.

Have you ever been surprised at how much sand comes home with you after a trip to the beach? Don't worry: there's plenty more where that came from. It's been estimated that if the sand on all of Earth's beaches and deserts was combined, there'd be 700 trillion cubic metres of it. This equates to roughly five sextillion grains of sand! Taking into account that this is only a rough estimate and could be off by a factor of two, this means there are between 2.5 sextillion and ten sextillion grains of sand in the world.

How does that compare with the number of stars in the universe? In our Milky Way Galaxy, we've estimated that there are between 100 and 400 billion stars. And in the observable universe, we estimate that there could be between 100 billion and 500 billion galaxies. If you multiply the number of stars by the number of galaxies, you end up with a range.

The total number of stars in the observable universe could be anywhere from ten sextillion to 200 sextillion.

Ten sextillion looks like this:
10,000,000,000,000,000,000,000
That is a one followed by 22 zeros.

Two hundred sextillion looks like this:
200,000,000,000,000,000,000,000.

Interestingly, the smaller estimated number of stars matches the larger estimate for the number of grains of sand. So there's a chance that there are as many stars in the universe as there are grains of sand in the world.

What's more likely, though, is that there are a mind-blowing five to ten times more stars in the observable universe than there are grains of sand in the world. Something to remember next time you brush yourself down after visiting the beach.

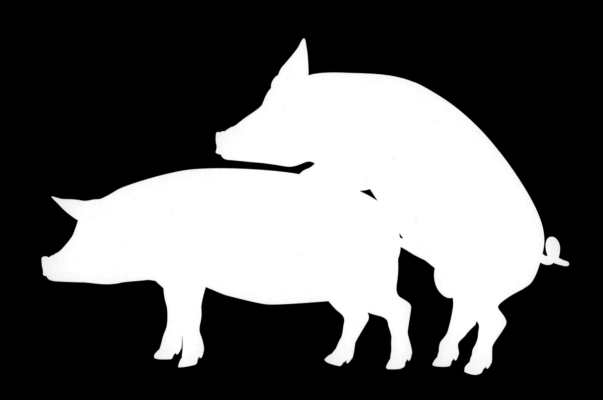

Pigs' orgasms can last for up to 30 minutes.

When people are asked what they'd choose if they could be reincarnated as any animal, the most popular responses usually involve eagles, lions, dogs, cats, dolphins. For watchers of certain TV shows there's also the odd honey badger. One animal that rarely makes the list is the pig. Maybe that would change if more people were to learn that male pigs don't measure the duration of their ejaculation in seconds, they measure them in minutes. And we're not talking one or two minutes, but up to 30!

While, on average, male pigs tend to end their sweet, sweet lovemaking with a mere 4- to 5-minute-long ejaculation, this can often extend past the half-hour mark. We'll never know for sure exactly what sensations a pig is experiencing during those minutes. But thanks to some brave scientists, we do have some inkling.

These intrepid researchers have taken matters into their own gloved hands to further investigate the male pig's sexual prowess by coaxing the sperm from the animal themselves.

They came up with some interesting findings. The main one was the confirmation that the mind-blowing 30-minute ejaculation period was definitely not a rumour. And the second conclusion was that it's a bad idea to interrupt the subject before he's finished. Those researchers must have found that out at their peril – frankly, this puts a fresh spin on the phrase, 'rough day at work'.

The implication here is that the pig was – as predicted – having a rather good time.

The Spanish flu infected 500 million people around the world.

Imagine a pandemic infecting one-third of the world's population and in which between 3% and 6% of all people on the planet perished. Sounds terrifying. The internet would go into meltdown as the world lost its collective mind. Well, a hundred years ago, this is what happened – obviously, minus the internet part …

The Spanish flu pandemic of 1918 killed between 50 and 100 million people across the globe, going down as the worst medical wipeout in history.

It was first observed in Europe, the United States and Asia before quickly spreading around the world, extending to the remote Pacific Islands and the Arctic. And somehow this pandemic managed to spread far and wide without the aid of commercial air travel!

To start with, the effects of the 1918 pandemic were generally mild. Those infected experienced chills, fever and fatigue but usually recovered after several days. Later that year, a second, deadlier and more infectious version of the flu appeared. Days and sometimes hours after developing symptoms, the skin of the sufferer turned blue and their lungs filled with fluid, causing them to suffocate to death.

Spanish flu killed more people in 24 weeks than AIDS killed in 24 years, and it took more lives in just one year than the Black Death did in an entire century!

The flu is a highly contagious virus that attacks the respiratory system. When an infected person coughs, sneezes or talks, the virus can be inhaled by anyone nearby. Please, next time you feel a cough or a sneeze coming on, cover your mouth!

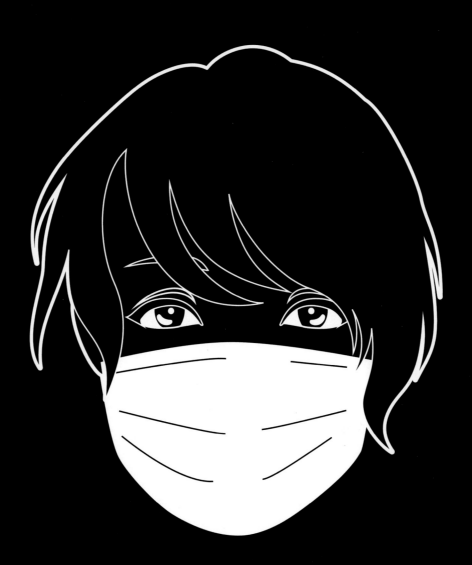

JANUARY	FEBRUARY	MARCH	APRIL	MAY	JUNE
The Big Bang		Milky Way Galaxy forms			Peak of star formation

JULY	AUGUST	SEPTEMBER	OCTOBER	NOVEMBER
Dark energy expands Universe	Our solar system forms	Earth forms and single cell life	Oxygen from photosynthesis	Complex cells

The 13.8 billion year history of the universe scaled down to a single year, where the Big Bang is 1 January and right now is midnight 12 months later.

DECEMBER

1	2	3	4	5 Multicellular life	6	7
8	9	10	11	12	13	14
15	16	17 Cambrian Explosion	18	19	20 Life leaves ocean for land	21
22	23	24	25 Dinosaurs ruled the Earth	26 First flowers evolve	27	28
29	30 Extinction of dinosaurs	31 Human evolution				

00:00	Apes and monkeys split
20:00	Humans and chimpanzees split
21:25	Humans walk upright
22:30	Human brain size begins tripling
23:52	Modern humans evolve
23:56	Human migration around the globe

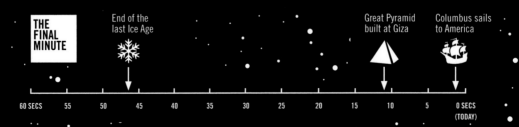

THE FINAL MINUTE

End of the last Ice Age

Great Pyramid built at Giza

Columbus sails to America

60 SECS   55   50   45   40   35   30   25   20   15   10   5   0 SECS (TODAY)

If you shrank the universe's 13.8 billion years into 12 months, civilisation would account for the last 14 seconds.

Putting our history as a species into cosmic perspective isn't easy. Humans tend to look at where we currently are as a civilisation and assume we've come a very long way over a very long time. Yet this isn't quite the case. Undeniably, we've accomplished a lot, but this has been done in a short space of time. Well … short in cosmic terms, anyway.

Although our ancestors have been around for something like six million years, today's modern form of humans evolved about 200,000 years ago. Civilisation as we know it is roughly 6000 years old, and our industrial revolution only started in the 1800s! Considering that the Earth itself is 4.5 billion years old and the universe 13.8 billion, we're really just babies — not even that.

One way to get a sense of cosmology, evolution and written history is to view it all in a cosmic calendar. Condense 13.8 billion years into just 12 months: one day is equal to 40 million years and a month to more than one billion years. How it works is that the Big Bang took place on 1 January at midnight and present time sits at midnight on 31 December. Now take a look at where we humans feature in the year.

At this scale, humans didn't appear until the very last day of the year, and our civilisation only accounts for the last few seconds of it!

Everything written about us in history books happened in those last few seconds.

Behind the curtain.

Hopefully you have enjoyed reading this book and have had your mind well and truly blown.

In order to research this book I ventured down a wonderful rabbit hole of books, websites, blogs, podcasts, YouTube, television and radio shows, live appearances and many conversations with incredibly smart people who have a much better understanding of the universe than I do.

Here's just a few of them:

BOOKS

Cosmos by Carl Sagan

A Universe From Nothing by Lawrence Krauss

Welcome To The Universe by J. Richard Gott, Neil deGrasse Tyson and Michael A. Strauss

Space Is Cool As Fuck by Kate Howells & Friends

WEBSITES

iflscience.com

reddit.com

nationalgeographic.com

livescience.com

howstuffworks.com

space.com

nasa.gov

universetoday.com

wired.com

isstracker.com

futurism.com

sciencedaily.com

nature.com

BLOGS

blogs.scientificamerican.com
reddit.com/r/science
newscientist.com

PODCASTS

The Infinite Money Cage
The Joe Rogan Experience
BBC Discovery
BBC In Our Time: Science
BBC In Our Time: Philosophy
Radiolab
Sean Carroll's Mindscape
The Tim Ferris Show
Stuff To Blow Your Mind
Psychedelic Salon
Under The Skin

YOUTUBE

Vsauce
CGP Grey
Big Think
Kurzgesagt – In A Nutshell
The Royal Institution

TELEVISION

Cosmos
Cosmos: A Spacetime Odyssey
Human Universe
Forces Of Nature
Life On Earth
Planet Earth
Blue Planet I & II
One Strange Rock
Our Planet

RADIO

StarTalk Radio

About the author.

Dan is a designer, illustrator and writer who runs a design agency, Studio Marshall (studiomarshall.com.au). For over 20 years he has worked with a diverse group of clients including Facebook, Sydney Opera House, The Australian Museum, One Laptop Per Child, Coca Cola and The Hunger Project.

Mind Blown was born from Dan's passion for graphic design, communicating information visually and his deep curiosity for the incredibly strange place that is our universe.